Al

Make Physics Great Again

America has failed

German original edition:
Einsteins Albtraum – Der Aufstieg Amerikas und der Niedergang der Physik

Make Physics Great Again - America has failed

ISBN: 9798372924765

Independently published

Contents

Preface

Why Europe and America are two different planets

Western civilization dominates the planet but is hardly heading smoothly and safely into the future. Further development will be shaped by science, and in the long run by fundamental science, in particular. To understand the current state of science, one must honestly examine its historical development. When it comes to science, we use the mind, which is influenced by traditions and cultures that form the human brain from childhood. It is these thinking traditions that are the subject of this book.

Modern science began about 400 years ago with the Enlightenment and underwent an extraordinary surge with the technological development of the late 19th century. This flourishing of the natural sciences originated in physics, whose insights into natural laws are among the greatest achievements ever accomplished by the human mind. The underlying culture of thinking originated in Europe, not coincidentally the continent that had dominated the world militarily and politically for centuries. In the early 20th century, by no later than the end of World War II, America had emerged as the leading power, and also not by coincidence, became the center of modern natural science.

Though seldom addressed by historians, this process was accompanied by a disruption of scientific culture. While the European natural-philosophical tradition of research focused on the fundamental laws of nature and pursued the question of "what holds the world together in its inmost folds," the technologically inclined culture of the New World was dominated by a desire to realize great visionary projects such as the atomic bomb and the moon landing, even if such projects had hitherto seemed unthinkable. While these projects may represent mankind's greatest technological achievements, they are not our greatest intellectual accomplishments. Whereas the technological-inventive element went hand in hand with fundamental research in the European tradition, the element of natural-philosophical reflection has been missing from U.S. scientific practice for nearly 100 years. This has obvious cultural roots. To put it bluntly, Americans don't like to think thoroughly.

This statement should be qualified in a number of respects. If one were to compare the living conditions of 100 years ago with those of today, no one could deny America's crucial contribution to the huge strides taken in civilization on the basis of that country's innovations, economic power, and values. Human history is replete with suffering and disasters; however, all the legitimate criticism of the U.S. notwithstanding, things could have been far worse in the previous century. If one is interested in the survival of civilization, it is necessary to analyze the mindset with which people explored the fundamental laws of nature. And it is obvious that there are glaring differences between Europe and America in this respect.

WHY ONE SHOULD DISTINGUISH WESTERN FROM WESTERN

We should not think about the abovementioned differences in strictly geographical terms since scientists enjoy a high degree of mobility and conduct research in many countries. Nor should we categorize individuals, as there will always be a number of exceptions to the rule; rather, we should consider the tradition of thought that prevailed in science at a particular time. The European roots of fundamental research in physics began to wither in the 1930s, and U.S. culture spread throughout Europe and beyond – what is today commonly referred to as "Western" culture. This indiscriminate term is one of the reasons why the long-term consequences of superficial thinking and the associated downsides of technology are often blamed on science as a whole, frequently leading to a general technophobia. Therefore, in order to understand the apparent crisis of "Western" civilization, it is imperative to study its precursor, the European research tradition.

The European culture of physics is rooted in the philosophy of the preeminent thinkers of ancient Greece. Later, physics became tremendously successful due to Galilei's empirical method. In America, on the other hand, physicists completely abandoned philosophy. It is true that certain philosophical arguments amounted to little more than a massaging of terminology, and it is not surprising that they never took hold in the New World. But since the country was completely devoid of a philosophical tradition, the theories developed in America often had roots that were too shallow.

This is not to be dismissive of theoretical physics in general. Specifically, no one should suggest the malicious distortion that the arguments put forward here have anything to do with "German physics". This was a group of experimentalists active in the 1920s who simply did not understand relativity and quantum theory and later tried to promote their careers in the Nazi regime by polemicizing against Einstein. Rather, it was Albert Einstein who symbolized the European culture of thinking and was a victim of the break with the physics tradition we are talking about.

Like no one else, Einstein had derived his insights from fundamental principles before he left for America, disgusted by the anti-Semitism. He was uninterested in the particle collider physics being pioneered there; conversely, his theoretical genius was barely appreciated, let alone used to reestablish the European tradition of physics, even though many others had turned their backs on the atrocities of the Nazis and the war in Old Europe. Indeed, Einstein was increasingly marginalized at Princeton, while the leading U.S. physicists pursued problems quite different from the fundamental questions Einstein pondered over throughout his life.

A NIGHTMARE IN SLOW MOTION

Tragically, with his letter to President Roosevelt in August 1939, Einstein also contributed to the development of the atomic bomb that finalized the transition of power from Europe to America. This weapon was the iconic symbol of a process that permanently shifted the focus of fundamental research from individual thinking to big science enterprises. American theorists believed that their success in having built the bomb qualified them as experts in fundamental physics. Alas, this was not the case. The postwar scientific supremacy of the U.S. was mainly a concomitant of its military and political power, while the bulk of the fundamental questions of the 1930s remained unsolved.

There is no doubt that the U.S. is still the dominant empire today. The history of how it achieved this position[1] is therefore worthy of closer inspection and will reveal what the consequences were for the scientific tradition. Modern science originated in Europe, and therefore I can offer little help to critics who label such an analysis as "Eurocentric". While conceding that my own view as a scientist has been shaped by the European tradition, I also admit that my views may be not entirely balanced. However, since contemporary physics has almost completely lost its roots, a healthy dose of counterpoise does not seem inappropriate to me.

> I can promise to be sincere but not to be impartial.
>
> - *Johann Wolfgang von Goethe*

The perspective I take when examining historical development is shaped by physics. Dealing with fundamental laws of nature is only a part of science, of course. But there are many unanswered questions in related disciplines: Why, for example, does the genetic code that underlies all terrestrial life consist of those particular 20–22 amino acids? One rarely hears such issues discussed in the current research culture. When I speak of science, I am often referring, in the strict sense, to elementary physics, although there is much evidence suggesting that similar patterns exist in the neighboring disciplines. Given the general traditions of thought, which can easily be identified, this would be unsurprising in any case.

THINKING WITHOUT DOMINANCE

To be clear upfront: while it does not mince words, this analysis is not "anti-American." There is hardly an empire in history without its dark side, but American virtues have certainly, on balance, advanced civilization. Perhaps the best example of courage, drive, and optimism, without too much "European" angst are the Wright brothers. Although a theorist had "proved" that a heavy body could not permanently hover in air, they simply built an aircraft that flew. Making the impossible possible is what defines the American dream to this day.

Nonetheless, both tackling problems and reflection are necessary for the sustainable development of our civilization. In Europe, people strove for insight rather than utility, for knowledge rather than power; discovery topped invention, and truth was valued more than mere success. The goal was to explain phenomena, not just describe them, and therefore theoretical understanding prevailed over practical application, unification was preferred to specialization. European scientists relied more on general principles than on abstract calculations. Generally speaking, they were more skeptical, but also humbler than their optimistic and sometimes conceited colleagues in America.

The American is kind, self-confident, optimistic and - unenvious. The European, on the other hand, more critical, more conscious, less kind-hearted and helpful, more demanding in his distractions... usually more or less pessimist.[2] - *Albert Einstein*

If one looks at the globalized world of today, these national categories can hardly be applied any longer; yet there is a crisis of fundamental physics permeating institutions everywhere that cannot be overlooked.[3] Its causes can only be understood if one considers the prevailing way of thinking, which essentially originated in America.

While the differences in terms of mentality and traditions are obvious, it is worth taking a thorough look at history, which is the focus of the book. It is clear that the different approaches are not limited to the natural sciences; in order to gain an overall understanding it will therefore be beneficial first to consider European and American culture in the fields of education, politics, and economics. These fields will also play a major role when, at the end of the book, we try to assess the implications of our findings for civilization; particularly since the cultural-historical consequences of this general crisis of thought appear not to be limited to physics.

Munich, January 2023

Part I: The Country without Culture

The attempt to unite wisdom and power was rarely successful, and if so, only for a short time.[4]
- *Albert Einstein.*

Chapter 1
Education is not a Value
Wisdom is not appreciated

Try to google "list of Greek ...", and the all-wise search engine will suggest adding the word "philosophers." This may by unsurprising for the cradle of European culture, but the corresponding query "list of American ..." first comes up with aircraft carriers and actors, before yielding a few hits for philosophers in the New World; not that the latter and their work should be underestimated. As far as the 19th century is concerned, Charles Sanders Pierce, above all, should be mentioned. As the founder of philosophical pragmatism, he rationalized the American way of thinking: the value of an idea is measured by its possible consequences. In society, however, philosophy was never held in the same esteem as it was in Europe, and many Americans would probably summarize pragmatism with the words: You do not need philosophy.

In fact, philosophy promises few short-term benefits and has therefore had a hard time from the outset in a culture dominated by success. Not much value was attached to general education as the main focus was on achieving results. By contrast, in Europe, the pursuit of knowledge has been entrenched in institutions for centuries, and education regarded as a value. This was the origin of the European success in science.

The Declaration of Independence in 1776 expressed, inter alia, a turning away of Americans from European values. We do not need you any longer – and we are getting along just fine. The Declaration guaranteed rights to *Life, Liberty and the Pursuit of Happiness*. No mention of education. No tradition emerged that would value the acquisition of knowledge. Why should it? In America, practical skills were needed. Eventually, the pioneering spirit of the immigrants and a flourishing culture of invention would take hold of the continent by the late 19th century. With English as the common language, a knowledge of foreign tongues was unnecessary for all practical purposes.

VERTICAL MOBILITY

Societies thrive because of the opportunities they can offer their members. Europe, too, owed its development to the fact that one could climb the social ladder through education and knowledge. Carl Friedrich Gauss, the preeminent mathematician of the 19th century, was the son of a craftsman. Johannes Kepler, who first described planetary orbits, was born into a poor family, benefited from the compulsory education the reformed Duchy of Württemberg in southern Germany granted to its children around 1580. It was the *Gymnasium* and the university that led him to the findings that were later to revolutionize the world.

By contrast, what became known as the American dream were the remarkable conditions in the U.S. at the beginning of the 20th century. With diligence, courage and entrepreneurship, one could reach the highest social status, which, in contrast to Europe, was primarily defined by wealth. Academic, let alone aristocratic titles and formal positions, had little meaning. This could happen in Europe as well.

Michael Faraday, a bookbinder apprentice with no formal education, studied all the scientific papers he could lay his hands on, thus becoming one of the most important physicists of the 19th century. Yet in Europe, education remained the yardstick that measured a career: the apprentice who becomes a famous professor. In America, on the other hand, success was defined by the proverbial rise from dishwasher to millionaire.

Whatever prerequisite (money, rich parents...) you needed for social success, education, let alone philosophical education, was not in the list. To this day, it is therefore considered a private matter and not something that guarantees the long-term survival of the state. Although much cutting-edge research is taking place in the U.S., the country's current educational system is in rather a dismal state – for a variety of reasons, of course.

The idea of a general educational canon[I] that makes one a mature personality is rather alien to Americans. Patriotism, on the other hand, is considered important. One is supposed to learn things about one's own country in the first place.[5] Moreover, Americans used to regard education as training; it had to be useful first and foremost. In science, this led to a focus on technical applications; people were more interested in the practical impact and less in deep understanding. When Einstein's General Theory of Relativity was spectacularly confirmed in 1919, the implications for space–time and for the universe were not a major issue. Rather, the *New York Times* dryly commented: "Einstein Theory Triumphs – Stars Not Where They Seemed – But Nobody Need Worry."

[I] Bill Bryson's *A Short History of nearly Everything* (2003) is an attempt in this direction.

PHILOSOPHY, THE SKILL OF THE UNEMPLOYED

We need to analyze how much philosophy is (under)appreciated in the U.S. because the philosophical foundations of science, let alone the corresponding way of thinking, never found their way into the country's physics institutions. In addition, European physicists sometimes heavily criticized schools of thought that created linguistic confusion rather than real insight. Einstein once called the writings of Karl Jaspers "drivel of a drunkard."

> Just look at today's philosophers, at Schelling, Hegel... and the like, doesn't your hair stand on end when reading such definitions?
>
> – Carl Friedrich Gauss

For Europeans, philosophy unfolded as a necessity. They saw themselves as natural philosophers in the sense that they were concerned with questions of space, time, and matter and wanted to understand the elementary laws of nature that governed these phenomena. The principle of sufficient reason, articulated by the 17th-century rationalist philosopher Gottfried Wilhelm Leibniz, was considered self-evident: everything must have a reason or a cause. This tradition of searching for the origin of natural laws certainly numbered among its adherents Goethe and Kant, who also considered themselves scientists, as well as René Descartes and the English natural philosophers George Berkeley and John Michell.

However, hardly anyone in America was enamored with these pioneers; not even people who had achieved success in science and technology. What was important was not the truth that nature might offer, but how things work. Thus, to the present day, philosophy has never achieved a special status in American culture. If nothing else comes to mind, think about the proudly anti-intellectual presidents such as Reagan, Bush Jr., and Trump. Imagine the faces of their supporters if they had quoted a philosopher at their election rallies.

The best philosopher is Jesus Christ.
– *George W. Bush*

All this is not to deny that American culture, freed from considerable European ballast, has led to great success. Let us compare two iconic figures.

FORESIGHT AND DEPTH

As role models of the traditions of thought in America and Europe, one can examine two researchers who were considered the leading scientific minds of their continents and who were almost contemporaries, namely Benjamin Franklin (1706–1790) and Leonhard Euler (1707–1783). However, their lives could hardly have been more different. Born in Boston as the son of a soap and candle maker, Benjamin Franklin learned the printing craft and steadily made his way up the economic ladder. Being also a successful writer, he acquired broad knowledge and even planned to found a scholarly society. However, he always focused on practical problems such as animal breeding, cultivation of crops, land surveying, and fire prevention.

At the mature age of 42, he turned to purely scientific topics such as electricity, which was still little researched back then. He acquired worldwide fame for inventing the lightning rod, undoubtedly of enormous importance at a time when cities were regularly devastated by fires. Political skill was also one of Franklin's many talents. He assumed public positions in the emerging independence campaign and was soon seen as a popular leader. Ultimately, he played a significant role in the War of Independence against the English motherland. While on a diplomatic mission to Paris, he succeeded in persuading France to enter the war on the side of the breakaway colonies, which ultimately led to the recognition of the United States of America in the Peace Treaty of Paris in 1783. Of course, this was only possible because he was already known in Europe as a famous inventor.

The life of Leonhard Euler, born in Basel, Switzerland, was quite different. Despite Leonhard's enthusiasm for mathematics, his father had intended to let him study theology. However, his son soon became a scientific star. The prodigy enrolled at university at the age of 13 and at age 16, he submitted a dissertation on the works of Newton and Descartes. Following this up with a second dissertation on sound propagation at the age of 19, he subsequently won the Paris Academy of Sciences prize 12 times and became arguably the most important mathematician of the 18th century; and not only because of his exceptional productivity (866 publications). While his findings on complex numbers, calculus, number theory, and geometry were groundbreaking, he also founded entire research fields in the applied sciences such as mechanics of fluids and rigid bodies. With an apparently unimpressive appearance and a personality that was not particularly endearing, he got into an argument with King Frederick II of Prussia and eventually accepted an invitation from Empress Catherine the Great to Petersburg, where he dedicated the rest of his life to science.

THINKERS HERE, DOERS THERE

If we talk about genius, depth and abstraction, Euler's work surpassed that of Franklin in every respect. Euler had prepared the mathematical ground for modern physics that later culminated in the Schrödinger equation of quantum mechanics. At the same time, his abstract approach also solved practical problems, just on a more advanced level. In addition, Carl Friedrich Gauss – who, like Franklin, was engaged in geodesy – delved much deeper into the subject and created the foundations of differential geometry, on which Einstein later built in his general theory of relativity.

Nevertheless, if we consider the practical benefit for mankind and the impact on history, Franklin was arguably more important than Euler. His aphorisms also testify to a wisdom of life that the aloof Euler had hardly acquired. All this leads us to believe that a coming together of the two ways of thinking exemplified by Euler and Franklin would have been of significant benefit to humanity. Yet, the tradition of profound philosophical thinking failed to appeal in America or what we call the Western world today. In particular, any echo of Socrates' insight, "I know that I know nothing," is rarely heard.

We know nothing, that is the first. Therefore, we should be very modest, this is the second. Not to claim that we do know we do not, this is the third. That is the attitude I would like to popularize. There is little hope for success.
– Karl Popper, Austrian–British philosopher of science

THE THING ITSELF MUST WORK

It is unsurprising that in a young state with many immigrants, erudition was less appreciated in America than in Europe. People were cooperative and helped each other, but the newcomers, although capable and skilled, first had to secure their economic survival. It is only natural that knowledge and education focused on these existential needs. For those early Americans, it was inconceivable that someone like Immanuel Kant in Königsberg would spend years of his life pondering over knowledge a priori and a posteriori, while never leaving his hometown. People simply did not have time to philosophize. In Europe, many who would afford themselves the luxury of thinking about all and sundry, for generations never had to worry about where their next meal was coming from. The French aristocrat Louis Victor de Broglie, for example, could thus dedicate himself to the profound study of unifying Einstein's findings on quanta and relativity[6] and discovered the wave nature of matter. He felt no need whatsoever to create immediate benefits, even though his discovery influenced our worldview (and numerous inventions) in a way few others have.

America recognizes no aristocracy
save those who work.
– Calvin Coolidge

Many people, however, did not have such means at their disposal and therefore sought their fortune in the New World. From 1820 to 1920, the population of the United States grew from 10 million to 106 million, about 36 million of whom came from Europe.[7] By the turn of the century, as many as 15 percent of the population were foreign-born. Along with those who had arrived a few generations earlier, these immigrants were to define American culture. Obviously, the new citizens had different values, education, and life goals than the Europeans. Hardly any of them were wealthy, and many were completely penniless; but they brought what was needed: the ability to work their way up economically, a virtue that in turn caused the country's extraordinary prosperity generations later.

The new world impressed me almost from the first day. The free, easygoing activity of young people, their spontaneous hospitality and helpfulness, the cheerful optimism that emanated everywhere, all this made me feel as if all burdens had been lifted from my shoulders.[8]
– *Werner Heisenberg, 1929*

PSYCHE OF THOSE WHO BURN THEIR BRIDGES

The emigrants were, however, a generation without roots who consciously abandoned the traditions of their ancestors, leaving the safe ground behind. The kind of conflict that such a decision usually entailed is illustrated in a scene from the film *Comedian Harmonists*: a member of an ensemble faces the prospect of leaving his lonely mother behind in Berlin. The moral question aside, emigration was a decision that said something about one's character: these people were looking forward to tomorrow, but were also heedless in a direct sense. It was the young, the strong, and the courageous who went to America, and it were these very qualities that resulted in the country ultimately becoming a world power. The self-deprecating Mark Twain expressed this in the witticism: "All you need in this life is ignorance and confidence, and then success is sure."

People poured into America because they longed for freedom; this was the basic reason. They sought independence from institutions. The few dominant religious beliefs in Europe turned into a variety of churches in America. Nonetheless, the immigrant's worldview was strongly influenced by Puritan Calvinism, which had become the predominant religious current in the New World after the Mayflower landed off Boston in 1620. The Pilgrim Fathers' belief of being an elected group with direct contact to the Creator later led to American exceptionalism, which – as expressed in the phrase "God's own Country" – continues to this day.[9] Paradoxically, therefore, the newly created nationality had a closer connection to religion, something we observe to this day. Just listen to the religiously soaked speeches of some American politicians – an unthinkable language in today's secular Europe.

GOD BLESS AMERICA

Not surprisingly, this religious imprint later morphed into an optimistic societal mindset that also found its way into scientific psychology. It was not for nothing that bestsellers such as Norman Vincent Peale*'s The Power of Positive Thinking and* Dale Carnegie's *How to Win Friends and Influence People* established a tradition of "you can do it"-"yes you can"-"believe in you" literature, which to this day promotes the American dream of the individual career. To what extent this is still in touch with reality is another question.

> We Americans devour eagerly any piece of writing that purports to tell us the secret of success in life.
> – *Franklin's biographer*

Accordingly, behavioral psychology – often based on empirical research – boomed in postwar America, while analytical psychology, founded by Sigmund Freud, was frequently demeaned as a relic of philosophical musings. However, not only was Freud a pioneer in his field, having discovered the interaction of psyche, dreams, and behavior, but there is no doubt that any intellectually satisfying analysis of psychological phenomena relies heavily on his findings.

However, as so often in America, people did not focus on theoretical understanding, but rather on practical benefits. Even successful psychoanalysts admit that understanding deep mental imprints does not automatically lead to therapeutic success, whereas the "superficial" behavioral therapies often yield amazing improvements for the patient. It is also noteworthy that individual psychology, with Freud, Jung, and Adler as its proponents, was dominant in Europe. Often based on single case histories, it had certain natural shortcomings in terms of comparability, reproducibility, and statistical methodology.

On the other hand, postwar experimental psychology produced spectacular results, such as Stanley Milgram's studies on authority (1968) or Solomon Ash's legendary conformity experiments (1953); even though some of its content had already been anticipated in Gustave Le Bon's *Psychology of the Masses* (1895). The European focus on individual problems slowly shifted to more practical applications of psychology, such as leading groups, influencing,[I] and, if necessary, manipulating them.[II]

Ironically, Sigmund Freud's nephew Edward Bernays, whose parents had immigrated to New York in 1892, played a major role here. Bernays is considered one of the founders of *public relations*, which at that time was still referred to by the more honest term *propaganda*. Among other activities, he advised President Wilson on how to win public support for taking America into World War I in 1917, leaving little doubt about his motives:

> If we understand the mechanism and motives of the group mind, is it not possible to control and regiment the masses according to our will without their knowing it?

The following statement of his is equally sobering:

> The conscious and intelligent manipulation of the organized habits and opinions of the masses is an important element in democratic society.[10]

Anyone who praises the virtues of Western democracies should at least have heard of Bernays. The mechanism of how collective opinions form certainly plays a major role in science, too. In any case, the European tradition was still focused on individuals, while science as a mass activity did not appear until large institutions were founded in America, which led to profound changes.

[I] As one of the pioneers of psychology, Kurt Lewin points out, there was a change of subject from individual psychology to leadership and management of groups and to communication. The American philosophers of the 20th century also focused on the areas of communication and semiotics.

[II] Whereby, of course, Germany with the hysteria orchestrated by the Nazis undoubtedly offered reason to deal with the phenomenon of group and mass psychology.

THE BULLY NATION

Plumbing the depths of understanding does not necessarily yield short-term benefits, and the one who skillfully and even shrewdly keeps his eye on his success often outplays, in terms of life goals, the deep thinker whose symphony of ideas remains unfinished. In the very long run, however, it was detrimental to American society that education and knowledge were regarded as mere tools for success and not as values per se. If one takes physical ability as a metaphor for a nation's strength, Europe was probably the equivalent of an imminent pensioner. America, on the other hand, was the stout youngster, lacking in wisdom to this day. The youngster has not aged well.

The process of growing up is nothing
but the individual civilization process.[11]
– *Norbert Elias*

Chapter 2
Science is not War
Where power harms knowledge

Human coexistence is the purpose of
culture, war is its opposite.
- *Carl Friedrich von Weizsäcker*

The way of thinking that has had such an impact on the development of modern physics can be found in many fields; and studying these parallels is necessary for an understanding of science. As noted earlier, scientific supremacy has almost always been accompanied by military dominance throughout history. On the one hand, countries with great scientific capabilities regularly used these capabilities for weapon technology, advancing their military. On the other hand, powerful countries also have more resources to invest in science. While these interrelations are obvious, the general mindset prevailing in American science deserves a closer look.

For its inhabitants, America is the hub of the universe, and, of course, *The Greatest Country in the World.*[12] While the media hyperventilates about domestic trifles it usually blinds out what should be of worldwide concern; yet countries on the other side of the globe are expected to obey American laws.

We don't know about anything that's
happening outside our country.
- *Michael Moore*

WHAT IS IMPORTANT HAPPENS HERE

For such a narcissist nation, it is also clear that America must be the Mecca of science; not the countries that are regarded as modern colonies. Such self-confidence has often led American physicists to constructing theories without considering the European precursors worth mentioning. The idea was that anything important could just as well be reinvented. Nonetheless, European theorists such as Albert Einstein, Paul Dirac, Erwin Schrödinger, and Nils Bohr were still the brightest minds in fundamental science, even after World War II.

However, even at that time it was unusual to translate European scientific journals into English, which is certainly one reason why so many old problems were forgotten.[I] Subsequently, the large number of inventions and the significant successes in applied science shifted the focus away from fundamental questions. Cutting the cord from Europe was part of the U.S. identity, and the desire to become completely independent was also felt in science.

For example, the controversial debate by the founders of quantum mechanics about its interpretation hardly interested anyone in America, as if it were a small war in old Europe that had been gladly left behind there.

[I] There are, however, wonderful initiatives such as the mathematician David Delphenich, who translated about 400 technical articles from European languages into English. Russian journals, which received attention during the Cold War, were also somewhat of an exception.

This entire country was built by Europeans who fled their homeland because they refused to endure the narrowness of the old continent, the eternal quarrels of small nations, the sudden switches from oppression to liberation, the frequent revolutions, and all the misery...[13]

- Enrico Fermi to Werner Heisenberg, 1939

Ironically, however, it is now the U.S. that is involved in almost every warlike conflict. Yet Americans have a hard time realizing that they are a part of history. They live in the present, in which they dominate the globe, and believe it is the most normal thing in the world.[I] With their large number of aircraft carriers, they control the oceans and can deploy troops almost anywhere; however, they also lead in modern fields such as space weapons and cyberactivity.

Empires have existed at all periods in human history. In most cases, they had the leading technologies at their disposal and, to the extent that these were useful for the military, always used them to expand their territory. The rise and fall of empires are therefore quite a natural phenomenon from a historical point of view. It would be downright strange if there were no dominant military-political power at present. The U.S. replaced the British Empire,[II] and before that, Spain was the leading power. There is, perhaps, a

---

[I] Noam Chomsky writes scathingly: " the political class [...] accepts it as normal and proper that the US should be a terrorist superpower, immune to law and civilized norms."

[II] The Anglo-Saxons who settled England also show some historical parallels to the immigrants in America; however, that discussion would go too far here.

particular parallel to be drawn between the U.S. and the Roman Empire. The latter was built on Greek culture, while the former was rooted in European culture, urging a comparison.

THE ANCIENT MIRROR IMAGE

Thales was the first to grasp the
principle of explaining the multiplicity of
phenomena... with the smallest
possible number of assumptions.[14]
- Erwin Schrödinger

Ionia, a cluster of Greek settlements on what is today Turkey's west coast, was the home of the most important philosophers around 500 BC. While they pondered over the building blocks of matter, the four elements (water, earth, fire, air), the Romans equipped their legions and built roads and catapults. As a modern parallel, the concepts of physics such as force, energy, mass, or even current and voltage, emerged in Europe. In America, on the other hand, powered flight, the assembly line and the machine gun[I] were invented. As in the case of America, the dominance enjoyed by the ancient Romans was based on their outstanding organization. Compared to ancient Greece, Rome had virtually no philosophers worth mentioning – also a mirror image situation. Finally, the emerging Roman Empire expanded thanks to its military technology; the same thing that can be observed with America since the beginning of the last century. But the similarities go beyond the military. Crucial to the cohesion of the Roman Empire were

---

[I] More specifically, by the American inventor Hiram Maxim in London, where he was allegedly advised " *If you want to make your fortune, invent something to help these fool Europeans kill each other more quickly*!" Whether real or well invented, there is certainly some truth in the statement.

the advanced technologies applied in daily life – infrastructure such as aqueducts and advanced new ways of building roads. The empire spread out across the Mediterranean, which ultimately became completely dominated by Rome. In a modern analogy, the U.S. regards practically all the world' oceans as *mare nostrum*. The U.S. dominance is also due to the modern digital traffic routes; the key infrastructure of the internet is controlled by America. In addition, the technological predominance of the NSA and the CIA enables an almost seamless monitoring and surveillance of global data traffic.[15]

All historical parallels aside, however, one thing becomes clear: Despite the wars of rivalry between Sparta and Athens or the brutal campaigns of Alexander the Great, the mindset of the Greeks was not imperial in the first place, but genuinely scientific, as Erwin Schrödinger points out in his book *Die Natur und die Griechen*. Europe also had its empires, but there was a remarkable international cooperation between scientists from the beginning of the 16th century, even though these activities were sometimes hampered by wars.

GOD'S OWN COUNTRY ON ITS MISSION

Ever since the founding fathers landed in the *Mayflower*, American culture has normalized the fact of being an empire. In the spirit of Calvinist exceptionalism, they considered themselves a God-chosen people to whom different rules applied. One cannot help but notice this way of thinking to the present day.

Accordingly, while nationalism and a sense of mission are strongly developed in the U.S., such lofty goals are not necessarily useful for scientific thinking. Nationalism was not very common among European scientists, except in times poisoned by war propaganda. They understood that scientific problems are not solved by nations.

Ruthless expansion is quite an accurate description of the history of the U.S., if one adopts the sober perspective of a Swiss scholar.[I] Its territory increased partly by cession, as with the French colonies, but also by wars of aggression, as against Mexico. Just as its sphere of influence grew steadily, it was later believed that science would likewise advance into new territory. With proper organization and funding, one can – so the thinking went – expand in any field, including politics and war.

Look, international law is... an important guide for other countries...[16]
- Max Uthoff, in the role of an American diplomat

MONOPOLIZING THE WORLD

It is true that the U.S. was smart enough to stay out of the conflicts in Europe for a long time. In the two devastating world wars at the beginning of the 20th century, they participated to the extent that it suited their interests[II] and took advantage of Europe's self-destruction to ultimately become the dominant power. By now, at the latest, the U.S. had started to like supremacy, not least due to the new possibilities of exporting goods. In the realm of science, this predominance also benefited the U.S., even though the flow

---

[I] Cf. the excellent book by Ganser (2023).

[II] During WWI, American banks had bought a considerable amount of English and French war bonds, which would have been lost in the case of a defeat of the Entente powers. In World War II, Stalin had long urged the Allies to invade France before 1944, but this was delayed until the Wehrmacht had been grinded by the Red Army (Ganser 2023).

was in opposite direction: all the talented people gathered in American universities.

World War I, with its numerous advances in weapon technology, was a watershed event that strengthened the ties between science and the military. By its nature, application-oriented physics in the U.S. had always been closer to military use.[17] Given the parallels of military and scientific success, it was tempting to conduct science the same way as war: with perfect organization and unlimited resources.

Later, the relationship between science and the military became even closer. People like Vannevar Bush, an engineer who advised President Roosevelt, is considered one of the founders of the military-industrial complex. In a 1944 cover story, *Time* magazine dubbed him the "General of Physics." Of course, an even more consequential reorientation of physics occurred at the end of World War II. Although based on European research and developed with the help of immigrants, the atomic bomb was built in America, thus ensuring the country's absolute military supremacy.

> Our national preeminence in the fields of applied research and technology should not blind us to the truth that, with respect to pure research… America has occupied a secondary place.[18] - *Vannevar Bush*

SOMETIMES ON THE SIDE OF MORALITY

Its fight against the Third Reich, as well as its opposition to the totalitarian regimes of Stalin's Soviet Union and Mao's China, gave the U.S. a morally superior role. Despite the ruthless actions in other countries, at least its own population was not sacrificed for the sake of an ideology. Compared to other parts of the world, the West stood for freedom, democracy, and human rights. U.S. economic dominance was so pronounced that other countries literally feared being pushed up against the wall by America's innovations.[19] At this time, U.S. exceptionalism, which had always existed, turned into exuberance; no doubt they saw themselves as the savior of the world.

This attitude led to outright craziness, as we have subsequently learned from audio recordings secretly made in the Oval Office by John F. Kennedy in 1962. Leading Pentagon strategists and generals openly argued for a preemptive thermonuclear strike against the Soviet Union and China, in order to eradicate the evil of communism once and for all. Kennedy was asked whether the death of 20 million Americans would be an acceptable collateral damage of the expected counterstrike. Kennedy kicked General Curtis LeMay, who had asked the question (LeMay would later be immortalized by Stanley Kubrick in the character of Dr. Strangelove), out of his office and angrily shouted: "and we call ourselves the human race!"[20]

The immediate dangers of the Cold War were overcome by the courageous actions of individuals, together with a degree of luck, but the idea of shaping the world – if necessary, by force – persisted. Probably no other language in the world uses the term *Nuke Them*. What an expression! By contrast, the idea of humans being one big family, as advanced by peace researcher Daniele Ganser, remains alien to many people.

Mankind must put an end to war, or war
will put an end to mankind.
- *John F. Kennedy*

SLOWLY DRIFTING INTO ABSURDITY

A common feature of politics and science is the hardly discernable shifting of standards that happens over generations. If the respective elites seal themselves off too much (power and intelligence sometimes facilitates that), there is the danger of a slow loss of reality that goes unnoticed within the group. All the agreed-upon values, even the original fundamental convictions, may erode and turn into their opposite.

No one with a sane mind who has followed U.S. global activities in recent decades can still recognize the spreading of freedom, democracy and human rights. Even the iron principles of the U.S. constitution have been trampled on since 2001 in a manner that would have made George Washington or Thomas Jefferson turn in their graves.

Analogously, anyone familiar with the convictions of the founding fathers of physics would come to a similar assessment when comparing the theoretical activities of recent decades with what was considered fundamental physics a century ago. The disconnect from reality happens almost imperceptibly, as soon as one loosens the anchoring in historical facts and focuses on the supposed wisdom and power of the present.

The idea of being the global leader in fundamental science spread quite naturally in America in the post-World War II era, alas without any objective justification. In today's politics, the dominant narrative is that only what is practiced in the U.S. is truly good governance. Many Americans are still convinced that all countries in the world should be made happy by the *American way of life*, and if necessary, even by force.[21] The old continent's reticence in the face of such a U.S. attitude is perceived as weakness. Accordingly, the U.S. regards itself as the bravest and strongest nation in science. It is true that America leads in space travel[I], computer science and controls the internet, which secures its global dominance to this day. However, one should not forget that these technologies were the consequence of a form of fundamental research that no longer exists.

In addition to contemplating how the U.S. became a world power, it is also interesting to examine its internal affairs, where violence was also always a common tool. Those who advanced the settlements in a westward direction, often did so with false promises to the Native Americans, ruthlessly wiping out their culture with superior technology. The keeping of slaves from Africa, and even trading in them, was long regarded as the most normal thing in the world. All of this did not exactly reflect the ideals of the Declaration of Independence, which states:

> We hold these truths to be self-evident, that all men are created equal, that they are endowed by their Creator with certain unalienable Rights...

COWBOYS - FREEDOM WITHOUT LIMITS

And indeed if it be the Design of
Providence to extirpate these Savages
in order to make room for Cultivators of
the Earth, it seems not improbable that
Rum may be the appointed means.[22]
- *Benjamin Franklin*

[I] In this respect, the Sputnik satellite developed by the Soviet Union in 1957 was an exception that would soon be corrected by the "race to the moon"; conversely, this also reflected the Soviet Union's strong imperial competition in the postwar period.

The culture of the Wild West still seems to have a significant influence on the American way of thinking. The state of anarchy, in which the law of the strongest prevails, is depicted as paradise with absolute freedom. Hollywood westerns used to play with emotions when, quite exceptionally, the good guy would prevail against the supposedly stronger party – but of course this never happened without gunfire. The notorious Second Amendment of the U.S. Constitution, continuously fought for by the gun lobby, supports the mentality that, if necessary, a problem can always be solved with violence. With their programs replete with brutality, television stations distribute a kind of a playbook for violence, while at the same time revealing a nipple ignites hysterical outrage. Physical love is taboo and physical violence the norm, if not a must, for film heroes.

> ...a cultural–political phenomenon. In these Hollywood films...conflicts are usually resolved by shooting, slashing and stabbing.[23]
>
> - *Wolfgang Bittner, German author*

Mark Twain's bon mot, "to a man with only a hammer, every problem looks like a nail," applies to today's America. No matter how complicated the ethnic, religious, or economic entanglements in the world's conflict regions may be, there is always a suitable bad guy who must be eliminated by *regime change.*

Anyone who observes these actions over a longer timespan will notice that such a foreign policy is short-sighted and superficial, even if the negative consequences only become visible much later and are covered up by the corporate media. What exactly does the West want to achieve?

Similar short-sightedness and superficiality also characterize large parts of science. There is a lack of vision as to the goal towards which technological progress should lead humankind, as well as a lack of reflection as to why the fundamental laws of nature have the form we observe.

The question of why is the mother of all natural sciences.
- *Arthur Schopenhauer*

THE CHANGE OF RESEARCH CLIMATE

Real advances in our civilization have only been possible in times of freedom of thought. To achieve such advances, people need to discuss unconventional ideas without repression or fear – something that can be traced back to the ancient city-states on the Greek islands.[I] Freedom, which according to Thomas Jefferson is the "first-born sister of science," is no longer in good shape in times of cancel culture and increasing censorship of public discourse. As soon as a topic is somehow politically charged, researchers prefer not to touch it.

U.S. soldiers stationed in 170 countries celebrate independence from major foreign power. - *Der Postillon (satirical news site) on the 4th of July*

While the Soviet Union was the country that persecuted dissidents in the previous century, today the West harasses people like Edward Snowden, Chelsea Manning, and Julian Assange, whose offense was to have reminded the wrong people about constitutional and international law. Yet the American public still regard themselves as freedom fighters

[I] Thales of Miletus, for example, held that one could understand the world if one only took the trouble to observe it properly (Schrödinger 1956, p. 72ff.).

in the tradition of the Founding Fathers, although this illusion has been shattered by reality for some time.

Analogous distortions are widespread – not only in America, of course – in science today: If one believes what one reads in the papers, legions of researchers are engaged in a heroic struggle to unravel the ultimate mysteries of nature. The journals are full of success stories on the research front, as if the present were by any means unique. But once one notices how far science has moved away from its ideal of a true search for fundamental laws, one realizes that science fails in much the same way as politics does.

Morals aside, the mindset with which scientific problems are tackled is very similar to how the military is wired: arm yourself with all kinds of equipment and apparatus, create institutions, form groups with a large number of people, and, if necessary, beat out the problem with brute force. In experiments, this is done with higher energies and larger telescopes, and in theory building by ever more daring designs, including the postulation of further dimensions of reality. It is just that science does not work like that. Instead, chance, attention, lucky circumstances, and serendipity play a role: often, it was a simple but creative linking of thoughts that led to the revolutions, not the campaign carried out with sheer force.

OMNIPOTENCE FIZZLES OUT

For many years, America was exhilarated by its capabilities, and in many respects, not without reason. Only a nation with many talents and huge economic power could accomplish things like the Manhattan Project and the Moon landing. In addition, it certainly required a spirit of vigor and enthusiasm characteristic of the U.S. These unique accomplishments, however, misled many into believing that America could achieve anything with the same means: almost infinite material resources, masses of skilled labor, and perfect organization.

This feeling of omnipotence, the idea of being capable of solving all problems by sheer force, was therefore not only prevalent in politics. A corresponding *brute force* thinking pervaded science, which had aspired to lofty goals, such as the fight against cancer.[I] For decades, the main research activity in America focused on trying out new cocktails of poison that would hopefully damage the tumor without killing the patient. These chemotherapeutic agents proved to be quite effective, yet they have not solved the real problem, which is tumorigenesis. By contrast, the fundamental work of biochemist and Nobel laureate Otto Warburg has gone largely unnoticed. According to him, the cause of cancer may lie in cell metabolism – a hypothesis that has only recently been taken up by Dr. Thomas Seyfried from Boston College.[24] Only the future will tell which approach delivered on its promises, but the difference between the cultures of the old and new worlds is obvious here, too.[II]

PARALLEL WAYS OF THINKING, NO GOAL IN MIND

Physicists approached the problems of elementary particles after the war with the same optimism and built ever-larger accelerators, a paradigm that continues to this day. I

[I] Even under Richard Nixon, otherwise not exactly an orphan, the USA set such goals serving humanity. This is rarely evident today.

[II] Conversely, one cannot deny that sometimes the grand visions of the general attack are successful, such as with the *Human Genome* Project.

will discuss later what this has led to, but again we note the change in mindset compared to earlier times. With similar optimism and a *manpower* of tens of thousands of theorists performing intricate calculations, contemporary science tries to solve the problems of the so-called string theory or supersymmetry; alas, in comparison with the beginning of the last century, with meager results.

It seems that the U.S. empire is not only overstretched politically, but above all lacks the intellectual foundations that would qualify it to rule the world responsibly. First and foremost, strategies would have to be developed to end the dangerous game of mutually assured destruction, which poses the risk of nuclear Armageddon for all mankind. Moreover, it should be obvious to any reasonable person that waging war in a technology-dependent civilization becomes increasingly irrational because the damage increasingly exceeds the benefits. For a sustainable global thinking in the interests of the well-bring of the planet and its inhabitants there is probably not only a lack of ideas and abilities, but also a lack of will, if not a total lack of understanding that this is a central task of our civilization.

Chapter 3

Crises, Bubbles, Crash

Symptoms of short-term thinking

At first glance, the prevailing economic situation and the state of science seem to have little in common. However, today's economic system and the way fundamental science is conducted display considerable parallels when we look at their way of thinking – to the extent that some concepts are completely analogous. To gain such a clear view, again we have to consider the historic development of science and economics.

In economics, there are usually relatively long phases of growth, which are typically followed by short, crisis-like events. These can be a plummeting stock market, a banking crisis, or even the collapse of the entire economic system. Growth is not always healthy. Time and again in history, there have been bubbles that, after an inconspicuous start, ended all the more abruptly in a crash, often accompanied by major devaluations and debt cancellations.

The philosopher Thomas Kuhn noted the exact same timeline of events in his work *The Structure of Scientific Revolutions*. Long phases of so-called "normal science" usually come to an end in sudden crisis called paradigm shift. As for the existing theoretical models, often no stone is left unturned. The textbook example is the Copernican revolution. For a long time, researchers had been blind to the signs of crisis in the so-called epicycle model that postulated excessively complicated planetary orbits in order to forcibly keep the Earth at the center of the universe. The dogmas of the model, such as that the heavenly trajectories had to consist of circles, were not questioned.

> If the Lord Almighty had consulted me
> before embarking upon his creation,
> I should have recommended
> something simpler.
> - *King Alfonso X of Castile on the Geocentric Model*

MANIA EVERYWHERE

However, there are also crises that occur on a much shorter time scale. Believe it or not, science can be as irrational as the infamous tulip mania in Holland around 1630, when tulip bulbs were traded for the equivalent value of a house. In 1903, for example, shortly after the discovery of X-rays, French physicist René Blondlot had convinced himself of having identified hitherto unknown "N-rays." Three hundred (!) papers were subsequently published before these rays turned out to be a collective self-deception of more than 100 scientists. Modern researchers should let that sink in.

Hints of an escalating crisis are often ignored – probably a characteristic of the human psyche that can be observed in many fields. But problems are also covered up in a very real way; just think of the banking crisis in 2008, which then saw the culprits being bailed out with taxpayer money. In a certain sense, this pile of money is analogous to the ever-higher energies particle physicists utilize in their collider experiments, while blinding out the fundamental contradictions of the model.

On the other hand, the inflation created by the FED and other central bank interventions can also be compared with the inflationary growth of free parameters in physics models. These poorly understood numbers are used to describe experiments that generate increasing numbers of particles. The complication that arose was the characteristic that made the geocentric model so impenetrable, but also finds a modern parallel in the opaque products of the financial industry. These products, too, help to put aside what has been produced in inflationary quantities. By analogy, "inventions" such as derivatives are a sign of instability of the system.

Just as the increased amount of circulating money in a period of inflation is less and less reflective of the value of real goods, the increase in arbitrary parameters when describing data no longer reflects the real value of observations. Economical debt thereby corresponds to those hoped-for observations that have not yet been made. In modern physics, much is postulated without direct evidence, just to keep the model running. With these ad hoc "explanations," the scientific enterprise muddles along, without "paying" to the problem in the form of real understanding.

DEBT AND GALACTIC SHADOW BANKS

To give an example, the inexplicably high rotation speeds at the edges of galaxies are "explained" by a huge amount of so-called "dark matter," while an anomaly regarding the expansion of the universe is ascribed to "dark energy." Ignoring for a moment the questionable methodology of such a "creation of credit," the sheer quantity of those dark substances is disconcerting – it is alleged that 95 percent of the universe consists of them.

The American cosmologist Michael Turner was once asked after a lecture whether such a "cosmology on credit" funded by these dark substances shouldn't be considered as risky as running a business with a corresponding amount of debt. His response, "In America, we're used to doing that!" amused the audience, yet carried more truth than many realized. If a model relies on so many untestable assumptions, one runs into a systemic risk similar to when being financed by excessive debt. In fact, we can observe similar mindsets – for example, prioritizing what keeps the system running over long-term stability. In the case of dark matter, one is reminded of the 18th century theory of phlogiston, a hypothetical substance postulated to explain fire phenomena. It turned out to be an illusion after the discovery of the real mechanism involving oxygen and oxygenation – another scientific revolution.

A weakness business and science have in common is the lack of instinct for unhealthy developments which, viewed soberly, are irrational. At present, ten trillion U.S. dollars per day are traded on the foreign exchange markets. Hypothetically, but based on the actual grain price, one could buy food for 30 years for the entire population of the Earth with that amount – without anyone having to work any longer. This is an equation that our economic system sets up, and its patent absurdity reveals a loss of reality that is entrenched in the superficial mindset of today's Western culture.

METHODOLOGICAL BLINDERS

Let us consider a similar absurdity in science. The production of data at the European Organization for Nuclear Research (CERN) has reached such an excessive level, that 99.99 percent (in the future even more) of results must be immediately deleted due to insufficient storage capacity.

Luckily, the same part of the data is considered irrelevant by the model assumptions.

Not only would any sane person get a feeling of resources being squandered here, but this selection of data – known as "triggering" – is, at its core, a flawed method. History shows that most significant discoveries have been made where they were unexpected; something that is ruled out here from the outset. The field has sealed itself off against further progress.

In real economics, such mismatches generate their own dynamics. In the above example of buying grain, the price would rise as soon as someone stocked up significant quantities. In the case of food, speculators probably shy away from doing this excessively because the hunger created would be bad public relations. Yet with sufficient funds, any strong player can increase the price by buying in a scarce good and then make profit by reselling it,[I] leading to still greater imbalance. There is simply too much money floating around.

One does not need a PhD in economics to realize that the entire system may become unstable as a result of this mismatch. Nonetheless, these inherent contradictions are not perceived by the public and hardly ever discussed by the media. It is a case of the money supply steadily growing, generating large private fortunes on the one hand and huge national debts on the other.

[I] As the legend goes, Thales of Miletus in ancient Greece, booked oil presses for small money in advance and then rented them out at a high price at harvest time.

RELENTLESS EXPONENTIAL FUNCTION

Such exponential growth inevitably leads to a crash. As soon as there are no longer sufficient tangible assets to match the large amount of money, this money, in practice, creates its own objects of speculation. It is not the intentional action of an individual, as in the hypothetical grain example, but the herd instinct of investors alone that generates the corresponding bubbles, which then burst as soon as the first try to cash in.

The focus on the present determines the course of events far more decisively than the lessons that could be learned from history. More generally speaking, Western thinking lacks long-term reflection on the state of the system. This phenomenon has been observed for about 100 years.

"Those who cannot remember the past
are doomed to repeat it."
-Georg Santayana[I]

Hardly anyone seems to be reflecting upon these matters; even among the experts there were only a few[25] who saw the financial crisis coming, which was then flooded with money printed by the central banks. The real, bigger crash is still to come. The collapse of a monetary system always leads to unmanageable inequities and is thus organized fraud. Lest we forget, in history this has often been accompanied by the collapse of states or even war. Compared to the above, what happens in science seems less dramatic, but a total loss of confidence in the integrity of insti-

---

[I] Spanish philosopher (1863-1952) who lived for a while in the United States.

tutions would have unpredictable consequences for the intellectual progress of mankind. The danger is not as small as it seems.

A positive feature of capitalism is that it removes from the market companies that fail to be competitive. Similarly, in a healthy scientific environment, theories that contradict observations are abandoned. However, this mechanism can become dysfunctional when formerly reputable theories are rescued with more and more auxiliary assumptions, as happens in the pre-crisis period described by Kuhn. Equally worrisome in the long run is the bailout of corporations that are "too big to fail," which poses a danger to overall stability. By alleviating the short-term pain, the risk of total collapse increases. Cheap money has now created many 'zombie' businesses that would no longer survive in a real market. One cannot help but come to the same conclusion with regard to certain scientific theories.[I]

After all, the time horizon politicians consider for their decisions is an election period – if not the most recent poll. The long-term danger to the financial system arose when Richard Nixon abandoned the gold standard in 1971. Paper ("fiat") money had lost the link to real values. Coincidentally, an analogous disconnect from reality – represented by experiment – could be observed in in physics around this time.

THE SHORT-WINDED CAPITALIST SYSTEM

While in science the sequence of long normality, short crisis described by Kuhn seems almost inevitable, one can argue whether such bubbles can be avoided in capitalism or

---

[I] For more details, see chap. 13.

arise by necessity. What is certain, however, is that in the West, the mindset that currently prevails in economics is too short-term and short-sighted. In the stock market, investors go for quick profits instead of long-term revenues. The pace of quarterly reports determines the strategy, not the investments in sustainable technologies. In addition, there is a fatal design flaw, namely that the resources of life – clean air, water, and the environment – are not reflected on the balance sheets. Similar patterns exist in science. In the day-to-day business of research reports, grant applications, and funding proposals, the real questions, which have been unsolved for 100 years, are usually forgotten.

However, the dangers of modern capitalism, even if not frequently discussed in public, should be clear to everyone. Yet these dangers are seldom associated with a particular tradition of thought and, of course, are not to be blamed on America in a narrow sense. Nevertheless, the rise of the U.S. to become the leading economic power after World War I had already led to a similar growth in both circulating money and debt at that time; and consequently, to excessive speculation. What followed the Great Depression of 1929 is well known. Economists nowadays believe they can prevent such a scenario by avoiding the mistake of money shortage, but at the time this only triggered the catastrophe. The common underlying problem persists.[I]

Clearly, crises such as that of 1929 are also related to traditions of thought. German philosopher of history Oswald Spengler was the first to discuss this linkage in his 1913

[I] However, the Glass-Steagall Act, which separated savings from investment accounts, certainly helped end the crisis. Today, obviously helpful would be a financial transaction tax.

work *The Decline of the West*.[I] When one learns that during the dissolution of the Roman Empire, the wheat harvest of entire countries was bought up for the sake of speculation, Spengler's claim that excessive monetary thinking is a sign of declining cultures seems to be corroborated.

Spengler's analysis is outdated in many respects. For example, he does not take into account how technology and globalization have accelerated civilization. But a connection between the shortcomings of an economic system and the underlying culture of thought certainly exists.

FIRE CIVIL SERVANTS, HIRE MANAGERS, AND EVERYTHING WILL BE FINE

Historically, there is a distinction between "Rhine" and "Anglo-Saxon" capitalism, which reflects in part the contrast between the European and American way of thinking. The latter is based on the ethics of success (bluntly speaking, success without ethics is also okay), which contrasts the continental European ethic of responsibility[26] that goes back to Immanuel Kant's categorical imperative: Act only in a way that one could make a law based on your exemplary behavior.

The Anglo-Saxon financial system is so dominant today that it is almost forgotten that there were functioning societies based on a stronger state with civil servants, such as Prussia in the 19th century. Prussia's economic system was certainly liberal, but that the state set the rules was never brought into question.[27] That was then, however. The U.S. tradition instead believes that if there is complete freedom

---

[I] A modern analysis of these ideas can be found in the book *Der lange Schatten Oswald Spenglers* by Max Otte (2018).

and an absence of rules, everything will take a turn for the better. Such a conviction is certainly rooted in the experience of the former colonies, which expanded practically without rules in the years after the Declaration of Independence.

Alien to the American way of thinking, for example, is the idea that private property or wealth may carry an obligation to contribute to welfare.[I] Instead, in today's international relations, property trumps everything, to the extent that sovereign states have to bow to the economic interests of individuals. Globalization plays a major role here, which is not necessarily such a bad thing. Yet it has led to a dilemma known in game theory: states compete in offering corporations the most attractive location, which makes the latter act ever more ruthlessly. Again, we encounter the idea that everything is better without rules.

It is interesting to note that Anglo-Saxon capitalism does not aim to abandon the poor and leave them to misery, but believes the problem can be solved through private initiative. In fact, there is a remarkable tradition of philanthropists like Rockefeller or Carnegie who followed the maxim "A rich man dies disgraced." One may object that these tycoons also wanted to buy power and a reputation for themselves, but the positive impact of their initiatives is undeniable.[II]

[I] As laid down, for example, in Article 14 of the German constitution.

[II] Not exactly in this tradition is George Soros, whose foundation has become a propaganda tool of US interventionist policy.

ALMS OR JUSTICE?

Bill Gates also transferred a large part of his fortune to a foundation. Dryly commenting "Being the richest in the cemetery doesn't matter to me," investor Warren Buffett also donated billions to that foundation. Bill Clinton's book *Giving* describes a variety of amazing initiatives that carry out humanitarian projects around the globe without any government funding. The rebuilding of Germany after World War II, which was largely financed by private donors from America, should also be remembered with gratitude.

The question remains, however, whether the state should really abstain from guaranteeing human rights. With all due respect for the philanthropists, it is contrary to European values to leave it to a donor's goodwill how and for whom the citizens' basic needs are provided. The 18th century social reformer Johann Heinrich Pestalozzi expressed this in the following robust way: "Charity means drowning one's rights in the shithole of grace."

Regarding science, the analogous question is how much fundamental science the government should finance. Ultimately, this cannot be left to the private sector either, because even if individuals are ready to fund important scientific projects, they will hardly make their selections without thinking about profit. There seems to be significant public funding for fundamental research in the U.S., but on closer inspection it is clear that this funding is often related to military projects.

JUSTICE, PUT TO THE TEST

The different traditions of thought in Europe and America are also revealed in the legal system. In European culture, law was supposed to realize justice as an abstract concept, often derived from theoretical considerations. In the Anglo-Saxon tradition, law primarily had to balance interests and risks, which is indispensable for any business activity. Accordingly, *case law* dominates, which is little more than an unsystematic collection of exemplary legal cases that have been decided. It would never have occurred to Americans to spend 20 years pondering over a thoroughly constructed legal system such as the German Civil Code (BGB).

Finally, the legal system reflects the psyche of people in a country of immigration. Where the established structures of old Europe were lacking, in practice it was simply the strongest that prevailed. Practical solutions prevented a slide into anarchy. It is typical of the U.S. legal system that one can be sued for almost anything. The system is based less on rules that may prevent adversity and more on sanctions after the fact.

Archaic juries that may impose capital punishment still exist in U.S. criminal law, which is quite incompatible with the idea of a civilized society. There is surprisingly little mention of this in Europe.

As much as I feel European, what I have noted above should not sound like a moral judgement. Obviously, however, deeply entrenched traditions of thought are revealed in very different areas. Taking into account that American society consisted largely of immigrants and their descendants, it is comprehensible from a psychological point of view, if not completely logical, that the described way of thinking emerged. The first people to arrive in the colonies had to endure hardships that can hardly be imagined today. No wonder that this society developed in a different way.

OTHER COUNTRIES, OTHER CUSTOMS

Business in the New World could not rely on governmental institutions as in Europe. However, private companies also behaved differently in Europe. The mere fact that a company had existed for centuries meant that it could be trusted as a business partner. If a merchant in Hamburg at the turn of the last century made a crooked deal, the damage to his reputation would have caused a much greater loss than what was dishonestly gained. To this day, however, the geographic mobility of people in the U.S. is much greater than in Europe. Unknown business partners entailed more risk, and adopting an appropriate strategy therefore became a necessity. The fact that fraud and dishonesty were sometimes almost acceptable in the U.S. cannot be condemned from a wider perspective. In terms of game theory, there were simply different conditions one had to come to terms with. An always trustworthy horse trader at the time of the Wild West would probably not have survived for long – at least not economically.

For those who arrived in America, life meant a lot of work, precariousness, and challenges. One needed pragmatic rules that were not too moralistic or deep in meaning, but also without too much of Europe's seriousness and bureaucracy. To secure your existence, you needed a job; not a sophisticated profession you felt a vocation for. On the other hand, all these factors formed a society that was more flexible and creative.

Since America drew its prosperity from economics and business, the corresponding way of thinking spread to other fields. The applied sciences that flourished in America had more in common with business than with the European tradition of thinkers and natural philosophers. It is fine when research results lead to useful technological applications, which means yielding real returns. But what about fundamental science? From the American point of view, an "investment" has to yield a "revenue" here, too, which is measured in the number of Nobel prizes or simply in the amount of research funds raised. Yet this has nothing to do with what Newton, Ampère, or Einstein were inspired by.

MONEY ON THE TABLE AND EVERYTHING IS POSSIBLE

For entrepreneurs, *think big* is considered the key to success. In U.S.-dominated science, however, this motto does not refer to big goals, but rather to large-scale projects, the managing of which America has always handled much better than Europe. These same skills were also used to strengthen the U.S. leadership role in science. We observe here a research philosophy based on material thinking: it is believed one can achieve anything if sufficient resources are provided; what was expensive must therefore be significant.

Money makes science fat and lazy.
- Fred Hoyle, British cosmologist

Alas, nature is not interested in research funds. The ever-growing particle accelerators have led to more and more particles, but not automatically – even if that was the belief – to new insights. The idea that anything can be achieved through unlimited use of money, power, and force, coupled with unwavering American optimism, can be observed in many areas of science.

Big science may destroy great science.
- Karl Popper

Although almost all major discoveries can be traced back to individuals, the role of the individual researcher becomes less significant in *big science* collaborations. The enterprise of science has shifted from the individual to the collective, and this significantly affects the way opinions are formed. Authority plays an important role, since large groups tend to strive for consensus. Again, we can see the parallels with economics and the rest of academia: the financial crash of 2008 was preceded by massive groupthink among experts. This also proves that the cognoscenti can be blind to long-term developments that erode the fundamentals of their own field.

In questions of science, the authority of
a thousand is not worth the humble
reasoning of a single individual.
- *Galileo Galilei*

Overall, today's science is dominated by U.S. culture in the same way the economy is. The common weakness is a lack of sustainability and basic reflections. The reasons appear to be buried in history. If one were to view fundamental physics as a business, we could say that the takeover by U.S. research traditions happened from 1930 to 1950; we will explore this in more detail in the following part.

Real progress would be best achieved by combining the European and American way of thinking. This probably holds true not only for physics, which is however the focus of the next chapters.

Part II
Rise and Crisis of the European Scientific Tradition

The supreme task of the physicist is to arrive at those universal elementary laws from which the cosmos can be built up by pure deduction.
– *Albert Einstein*

Chapter 4
Foundations
The big discoveries of the natural philosophers

Shortly after Christopher Columbus landed in America in 1492, Nicolaus Copernicus travelled from his native town Thorn (now in Poland) to Bologna in northern Italy with a view to studying law. Although he was invited to lecture on astronomy in Rome as early as 1500, he also started studying medicine in Padua and was regarded by his contemporaries as a distinguished physician. During the Renaissance, a spirit of enthusiasm and a thirst for knowledge rapidly spread throughout Europe, and numerous universities were founded. Copernicus was in the privileged position of having a wealthy uncle who made his studies possible, and from 1512 he could to pursue his astronomical ideas in a permanent position as canon of the monastery in Frauenburg. At the urging of the mathematician Rheticus, in 1543 he finally published his legendary work *De revolutionibus orbium coelestium.*[I] Although Copernicus is considered the founder of the geocentric worldview named after him, he took his basic idea from the Greek scholar Aristarchus of Samos. Greece is rightly considered the cradle of natural philosophy.

Natural science is thinking about the world in the way of the Greeks.[28]
- *Erwin Schrödinger*

[I] "On the Revolutions of the Heavenly Spheres".

Compared to Copernicus, Johannes Kepler, born in 1571, grew up in precarious conditions. It was only because education was held in high esteem in Weil – one of the free cities of Württemberg in southern Germany – that Kepler could attend elementary school for free, attend the grammar school in Maulbronn, and eventually study mathematics in Tübingen, where Michael Maestlin discovered his extraordinary talent. After painstakingly analyzing astronomical data, Kepler realized that planetary orbits were indeed ellipses, not circles. This was the decisive improvement of the Copernican model, which otherwise would never have gained acceptance. Physically slender and of fragile health, he was nevertheless driven by an insatiable curiosity from young age. If you ever happen to visit the Maulbronn monastery in Württemberg, while contemplating its impressively high windows, you will easily imagine how the young Kepler may have pondered there about the incidence of the sunrays.

To do astronomy is to read the
thoughts of God.
- Johannes Kepler

Today, the churches are considered the antagonists of science. Nonetheless, at that time, the clergy was the intellectual elite, and geniuses like Kepler could flourish in such monasteries. Later, in the midst of the Reformation wars, he held positions in Graz and Prague. One can only admire Kepler's perseverance in pursuing his ideas under such difficult conditions. The successful development of his third planetary law alone took him 10 years of frustration and errors in calculation before he eventually got to the breakthrough.

SINGLE MAVERICKS

Isaac Newton spent no less time on developing his law of gravitation. The first thoughts came to him when he had isolated himself in the countryside, while the plague was raging all over Europe. Newton, too, was one of those ingenious mavericks who were driven by an overwhelming curiosity. He was probably fascinated by the simplicity of his inverse-square law,[I] from which Kepler's elliptic orbits followed as a direct consequence. His discoveries were subsequently to lead him to the philosophical insight that "Truth is ever to be found in simplicity, and not in the multiplicity and confusion of things."

It is in the very nature of the pursuit of knowledge... to strive for simplicity and economy of the underlying hypotheses.[29]
- Albert Einstein

Galileo Galilei, who, as a character, perhaps came closest to the practical "American" way of thinking, also played an important role in convincing people of the new heliocentric world view. Although he held a chair of mathematics at the University of Pisa at the early age of 30, his obvious talent for handcrafting technical gadgets was no less important to him. Galilei was probably the first to properly understand the telescope, invented shortly before, and improved it significantly. This even proved to be economically useful: With the help of a telescope, the Doge of Venice could spot merchant ships appearing on the horizon and sell off goods before anyone else knew of their arrival.

[I] When doubling the distance, the force decreases to one forth, and so on. This idea however seems to go back to Robert Hooke.

More important for the progress of humankind, however, were Galileo's legendary observations of the phases of Venus and his discovery of the moons of Jupiter around 1610. Both phenomena were unmistakable evidence of Copernicus' theory. It is on the basis of these findings, together with his experiments on mechanics, that Galileo is considered the founder of the empirical method. After all, it is only empirical evidence that makes it possible to separate the wheat from the chaff among the philosophical reflections on the laws of nature. The empirical method was to lead to the triumph of science in subsequent centuries.

At the time when the first colonies were being founded in Virginia on the east coast of what is today the U.S., Galileo disseminated the new world view in his writings, while cheekily mocking the Pope, a frivolity that almost led to his being burned at the stake. Later, in 1687, Isaac Newton provided the theoretical justification for the new world view with his work *Philosophiae Naturalis Principia Mathematica* that soon was to become known throughout Europe.

Together with England, Germany and Italy, France was counted among the leading nations in science.[30] Charles Augustin de Coulomb realized that the structure of Newton's law of gravitation could also be observed with electric charges (which, by the way, had already been discovered by Thales of Miletus). The origin of this fascinating parallel of the two most important forces of physics is not understood to this day.

> For instance, nobody knows why the electric charge is always an integer multiple of that of the electron.[31]
>
> *- Emilio Segrè, discoverer of the antiproton*

PIONEERS OF KNOWLEDGE

The origins of modern science were observations of the starry sky, a pure form of human curiosity. Until the end of the 19th century,[I] however, astronomy was practiced mainly in Europe. The discovery of Uranus and Neptune had completed the observations of the solar system, while the existence of galaxies had already been suspected by Immanuel Kant. All this originated from a spirit of research that sought to comprehend the world without thinking of applications or even of profit.

Two things fill the mind with ever-increasing wonder and awe... the starry heavens above me and the moral law within me.
- *Immanuel Kant*

An influential thinker with this mindset was Alexander von Humboldt, pioneer of the holistic view of the environment, whose major work *Kosmos* sought to "grasp nature as one great whole, driven and animated by internal forces." Being a true polymath, he made significant contributions to physics, geology, mineralogy, botany, zoology, climatology, oceanography, astronomy, and chemistry. From 1799 to 1804, Humboldt undertook an expedition to Central and South America that made him the leading scientific authority of the time – a "new Aristotle."

After the pioneering work of Newton, who had remained without a proper successor in England, in continental Europe, Leibniz, Helmholtz, Euler, and Lagrange deepened the mathematical foundations of mechanics, their work

[I] Thus, the American Astronomical Society AAS was not founded until 1899.

constituting the basis for the further development of physics. Towards the end of the 18th century, the U.S. was founded, as mentioned, with the help of Benjamin Franklin, who, incidentally, shared Galileo's proclivity for satirical writings. The third president of the U.S. was Thomas Jefferson, who invited Humboldt, on his return to Europe, to a dinner in Washington in 1804. The statesman was impressed by Humboldt's erudition,[32] but particularly appreciated his knowledge of geography of South America and the corresponding maps, which were useful for Jefferson's expansion plans. By contrast, European maps pose a problem for some contemporary U.S. presidents.

SEARCHING FOR THE ELEMENTARY LAWS OF NATURE

The momentous discovery of the wave nature of light can be traced back to the Dutch physicist Christiaan Huygens (1629–1695), as well as to Thomas Young (1773–1829) and August Fresnel (1788–1827), all of whom were multitalented characters. However, it took until the 19th century for the second fundamental force of nature, electromagnetism, to be discovered. French theorists were at the forefront of this development: Charles-Augustin de Coulomb and, above all, André-Marie Ampère, who was also driven by an insatiable curiosity. In 1787, by the age of twelve, the child prodigy Ampère had already worked through all 20 volumes of the Encyclopedia by Diderot and d'Alembert, one of the main works of the Enlightenment, which contained practically all the knowledge of the time.

However, the link between electricity and magnetism became known only around 1820, through Hanns Christian

Ørstedt's[I] legendary observation of the deflection of a magnetic needle by an electric current. The sensational news from Copenhagen spread like wildfire. Ampère was the first to describe the phenomenon in theoretical form, while the abovementioned Michael Faraday revealed the systematics of the interrelation with a series of experiments, all of them meticulously documented.

After pioneering work by Wilhelm Weber,[33] who, together with Carl Friedrich Gauss, built the first telegraph in Göttingen, Scottish physicist James Clerk Maxwell crafted the theory of electrodynamics into a final form, revealing a far-reaching symmetry of the electric and magnetic fields. This was an unprecedented achievement of mathematical abstraction, which prepared the ground for even more sophisticated fields such as differential geometry. Particularly noteworthy is William Roman Hamilton from Dublin, who in 1843 invented a four-dimensional generalization of the complex numbers (quaternions), which Maxwell then used then to formulate his theory. What could be called cutting-edge mathematical research at the time, however, always had a direct link to the reality of physics.

Maxwell's equations, perhaps like few others, are fascinating in how they describe a variety of natural phenomena by the simplest mathematical structures. The celebrated summit of the theoretical developments was the proof of electromagnetic waves, provided by Heinrich Hertz in Karlsruhe in 1886. The existence of these waves had been predicted by Maxwell's equations, while their speed matched that of light. This was huge. Electromagnetism

---

[I] Italian physicist Gian Domenico Romagnosi had already published the same experiment in 1802, but it had gone unnoticed.

was united with optics, and the last doubts about the wave nature of light disappeared.

> In the beginning (if there was such a thing), God created Newton's laws of motion together with the necessary masses and forces. That is all; everything beyond this follows from the development of appropriate mathematical methods by means of deduction. What the 19th century has achieved on this basis must evoke the admiration of every sensing human.[34]
>
> - *Albert Einstein*

SEEKING EXPLANATIONS

Though almost forgotten from today's perspective, it is remarkable that the physicists of the 19th century were by no means satisfied with electrodynamics. Despite the formal analogy between Newton's and Coulomb's laws, there was a wide chasm separating the two theories. Maxwell, as well as many others such as the Irish physicist James MacCullagh, had tried to understand electromagnetic phenomena by mechanical analogues. Indeed, there is a striking similarity to the equations governing continuum mechanics that led people to assume that empty space was an elastic body called ether. It would be beyond the scope of this book to go into the history of ether theories[35] and to analyze when and why they were abandoned. The reason is not, as commonly assumed, incompatibility with Einstein's theory of relativity.[36] In any event, in the tradition of European physicists it was imperative to look for fundamental explanations, in this case by means of Newton's mechanics. The British super-authority of physics, Lord Kelvin, criticized electrodynamics throughout his life, arguing that it was not a theory that really explained phenomena.

Therefore, I cannot comprehend the electromagnetic theory of light... But I want to understand light as well as I can, without introducing things that we understand even less of.[37]
- Lord Kelvin

The term "ether" alone elicits a bored yawn from most contemporary physicists influenced by U.S. culture. Not only is this a sign of historical ignorance, but it also proves that people have bidden farewell to aspiring to a thorough understanding of natural phenomena. Of course, this cannot be blamed on America back then, where almost no theoretical physics was practiced at that time. Nevertheless, a problem was swept under the rug by arguing that the theory was successful and worked well, a view that is widespread today. On the other hand, the uncompromising search for fundamental laws of nature remains an element of the European tradition, which later gradually eroded.

While Einstein wanted to know if the universe had emerged inevitably from the laws of physics, his successors now want to declare that the laws of physics are inescapably determined by the way the universe looks.[38] – *David Lindley*

An example of someone who demonstrates that our discussion is not about nationality but thinking traditions is the American physicist Joshua Willard Gibbs (1839–1903), who made decisive contributions to thermodynamics. Since Sir Francis Bacon (1561–1626) had understood that heat had to do with particle motion, the foundations of the kinetic theory of gases had been developed by Robert

Mayer (1812–1878), James Prescott Joule (1818–1889), Rudolf Clausius (1822–1888), Gustav Kirchhoff (1824–1887), and the already mentioned Lord Kelvin (1824–1907). Gibbs, who had studied briefly in Europe, returned to a professorship at Yale in 1871; in the following years, he developed general results that are among the most important tools of modern thermodynamics. He was considered a genius a little disconnected from reality, and at the time his work was believed to have no practical significance. Tellingly, his dissertation at Yale University about an abstract mathematical problem had to be justified by an application in cogwheel technology before being accepted.[39]

TWO MILLENNIA, ONE PARADIGM

Thermodynamics as a whole is unthinkable without atomism, which is one of the most important insights ever arrived at by mankind. The idea of indivisible building blocks of nature goes back to the Greek philosophers Leucippus and Democritus, who seem to have been uniquely ahead of their time. The concept of atoms was elaborated in centuries of painstaking work into a model that today seems so self-evident that we are not even aware of its philosophical implications. John Dalton (1766–1844) and Amedeo Avogadro (1776–1856) deserve a mention as pioneers in this field, as well as Dimitri Mendeleev (1834–1907) and Lothar Meyer (1830–1895), the discoverers of the periodic table. However, it was not until the 20th century that a consistent formulation of atomic physics, that is, by means of quantum mechanics, took place.

Nothing happens at random, but everything for a reason and by necessity.- *Leucippus*

The reason for the extraordinary progress of science was the method *ratio et experientia,* established at the end of the 17th century, from which point we can draw a direct line to the triumph of technology in the second half of the 19th century.[40] Although the mechanization following the steam engine led to the first industrial revolution, the technological use of the findings about light and electricity occurred much later. The development of electrical engineering, in turn, led to fundamental new discoveries through experiments, such as the identification of the electron by J.J. Thomson in Cambridge in 1897.

In the history of physics, one can often observe such a co-evolution of theory and technology, which provide each other with new insights. The theoretical breakthroughs in the middle of the 19th century were followed, after a short delay, by a flourishing of the corresponding technology. This wave soon hit America, which at that time was recovering from the aftermath of the Civil War (1861–1865). At the same time, talented people from all over the world went to the U.S.

THE ELECTRIC ANTIPODES

After Werner von Siemens had discovered the dynamo-electric principle and developed the generator in 1866, the second, "electric" industrial revolution gained momentum. In the subsequent period, there were two great pioneers of electrical engineering working in America who better than anyone else demonstrate the European–American contrast regarding thinking traditions: the rivals Thomas Alva Edison and Nikola Tesla.

Edison, born in Milan, Ohio, in 1847, was the prototype of the American inventor. Having become a celebrity for his light bulb, he was familiar with literally every technology of the time. Edison developed groundbreaking innovations in electrical engineering, power generation and distribution, telecommunication, and audiovisual media. With an unmatched intuition for practical applications, he rushed to technological implementation before anyone else could. At the same time, he was also a successful entrepreneur who knew how to commercially exploit an invention. His contribution to the electric power grid of New York marked the beginning of the electrification of the industrialized world as a whole.

> Anything that won't sell, I don't want to invent. – *Thomas Alva Edison*

It is not too much to say that America's success was based on people like Edison. In 1997, *The New York Times* wrote that Edison's most important accomplishment was not the inventions themselves, but the invention of the invention industry. Edison, so the article states, was nothing less than the father of modern research and development. His biographer Paul Israel[41] particularly emphasized Edison's ability to think by analogy and to learn from failure. The wealth of ideas contained in his notebooks, however, was almost incomprehensible to an ordinary mortal.

Nikola Tesla was quite different. Born in 1856 near the border of Serbia and Croatia as the son of an Orthodox priest, his life was full of setbacks, illnesses, and economic failures. Certainly, Tesla would have had an easier time with a permanent position at a European university. Although his genius was probably evident early on – Tesla mastered eight languages – he dropped out of his engineering studies at the University of Graz in 1877 for unknown reasons. He tried jobs as an assistant teacher and telegraph technician before reaching New York in 1884 as one of many penniless immigrants. Such stories give one an idea of how that country opened up opportunities for talented people without formal qualifications. America was their aspiration for a better life.

THE ECCENTRIC GENIUS

Tesla took up employment in Edison's company, but quit after six months. He found his salary was too low and tried to create his own business. Yet, despite his legendary inventions in electrical engineering, he was often cheated of the fruits of his labor. Unlike Edison, Tesla never achieved lasting wealth through his product developments. With advancing age, his ideas became increasingly exotic, which is why his character is often described as bordering between genius and madness.

While Edison made a fortune with his inventions, Tesla invested what little money he had in his ideas. Although he became famous, some wealthy patrons had to cover his living expenses in his last years. Only in one, albeit technologically decisive point, did he beat his rival Edison: Tesla's system of transmitting energy by alternating current won a $100,000 award in a contest set up by the operators of the Niagara power plants in 1893. It was subsequently his technology that prevailed worldwide over Edison's direct current system. Tesla's extraordinary capabilities and his eccentric personality are the subject of numerous myths. Allegedly, the key idea for alternating current flashed into his mind in a park in Budapest in 1881, while he was reciting from Johann Wolfgang von Goethe's drama *Faust*. In purely intellectual terms, Tesla was undoubtedly superior to Edison.[1] Probably with some envy of Edison's economic

---

[1] Supposedly, the Nobel Committee intended to honor Edison and Tesla for

success, but nonetheless to the point, he wrote the following about Edison:[42]

> I was almost a sorry witness of his doings, knowing that just a little theory and calculation would have saved him 90 per cent of the labor. But he had a veritable contempt for book learning and mathematical knowledge, trusting himself entirely to his inventor's instinct and practical American sense...

DAWN OF A NEW AGE

The end of the 19th century saw an explosion of technology, which underpinned the emergence of the U.S. as a world power. Research flourished in the U.S.; however, unlike in Europe, it was focused on practical applications and inventions. Add American optimism, and one gets a sense of how people came to believe that, given such prosperity, the essential laws of nature must have already been discovered. All that was needed for a golden future was to further develop the technologies based on those laws. A celebrity who expressed this kind of enthusiasm was Albert Abraham Michelson, who in 1907 was to become the first American Nobel laureate in physics. His parents had immigrated to the U.S. when he was two years old; and during his career as a professor at the University of Chicago he became famous for his precision measurement of the speed of light.[I] However, he grossly misjudged the state of physics in its entirety in 1894:

> It seems probable that most of the grand underlying principles have been firmly established [...] future truths of

---

their accomplishments with a joint award, but Tesla refused to share the prize with his rival. Both remained empty-handed.

[I] These measurements were partly carried out in Potsdam in 1879-1883.

physical science are to be looked for in the sixth place of decimals.[43]

Short thereafter radioactivity was discovered in Europe, the consequences of which were soon to overturn the whole of classical physics. In early 1896, French physicist Henri Becquerel was stunned to observe that photographic plates in a dark drawer appeared to have been blackened by pieces of pitchblende (containing uranium) that he had stored in the drawer. Together with Marie Curie and her husband Pierre, with whom he shared the 1903 Nobel Prize, Becquerel examined the substance in more detail. Curie succeeded in separating the highly active element radium, for which she received the Nobel Prize in Chemistry in 1911. Almost at the same time as the discovery of radioactivity, Wilhelm Conrad Röntgen produced X-rays for the first time at the University of Würzburg. The legendary photograph that showed the bones of his wife's hand immediately travelled around the world.

It is remarkable how these discoveries were dealt with in America. In 1920, a Boston inventor filed a patent for a "pedoscope," an apparatus to be used in shoe outlets for checking by means of X-rays whether a foot would fit into a boot. For medical reasons, children, in particular, were advised to optimize their wear comfort in this way. Even more dangerous than X-rays was painting watch dials with radium, which yielded a nice glow at night. Many of these "radium girls," unfortunate painters who had contaminated themselves by pointing their brushes on their lips, later died of cancer. Yet these miraculous radioactive substances were even advertised as antibacterial toothpaste. Of course, these tragic events are not exclusively the fault of "American" thinking in science. One could also argue that in Europe there are too many reservations regarding new technologies, often in the form of "German Angst." However, this episode remains a historical spotlight on how new discoveries were regarded and commercialized in the USA, sometimes without sufficient reflection.

BOOM OF INVENTIONS

Science Finds, Industry Applies, Man Conforms! - *Motto of the Chicago World's Fair 1933*

The spirit of invention flourished in the U.S. as early as the 19th century and produced numerous technical gadgets. Samuel Morse, for example, independently developed the telegraph (which had already been built by Gauss and Weber in Göttingen in 1833) and, as a practical supplement, invented the alphabet named after him. The dynamics of progress in the New World were superior to Europe in many ways. Aviation is perhaps the best example of how the American habit of not always putting concerns first was superior to the reflective European approach.

Although "flying man" Otto Lilienthal had been an important precursor, the Wright brothers[44] achieved the first fully controlled[I] powered flight in 1903, bluntly disproving the claim of physics luminary Lord Kelvin, who a few years earlier had argued that a body with a density greater than air would never fly. The Wright brothers had probably never heard of Kelvin, and rightly so. They just built their plane. Similarly, the French mathematician Henri Poincaré had once "proved" that radio communication from Europe to America was impossible due to the curvature of the

---

[I] For this purpose, three different rotations, called *roll, pitch, yaw,* must be implemented.

Earth. Shortly thereafter, Guglielmo Marconi from Bologna, who in this respect had an "American" entrepreneurial spirit, simply did it.[I] Nonetheless, at that time the U.S. was still hopelessly behind in terms of fundamental physics research.

[I] Marconi benefited from the then unknown reflection of the waves from the ionosphere.

Chapter 5
Quakes and Tremors
How Einstein's revolutions came about

Born in 1879, Albert Einstein represents the European scientific culture like perhaps no other. Working as a lonely maverick, he not only had a unique overview of the fields of theoretical physics but also designed certain key experiments.[45] Rather disinterested in technology and its applications, he focused on the fundamental questions of nature and, through his special talents, delved deeper into the subject than most others. He considered intuition rather than computational skills as his strength. He may be called the "inventor" of the thought experiment and always tried to derive laws of nature from elementary principles.

I believe in intuition and inspiration. Imagination is more important than knowledge.
- *Albert Einstein*

At the time of his great findings, virtually no theoretical research was being carried out in the U.S., yet in Europe there was considerable technological development. Einstein's father Herrmann, for example, ran a company in Munich that installed electric street lighting, a high-tech enterprise at the time. Albert, on the other hand, was more fascinated by abstract things. The introverted kid, as he later recounted, had been profoundly impressed by the needle of a compass. Another key experience was finding a proof of the Pythagorean theorem at the age of 12. On the other hand, Albert did not particularly like his grammar school, the Luitpold-Gymnasium, which smacked of Prussian discipline. When his parents moved to northern Italy in 1895, he dropped out of school at the age of 15 and only later graduated from a small school in Switzerland. Around this time, he was already pondering what would happen if he were to move at the speed of light alongside a light wave.

THINKING TRUMPS CALCULATING

In his autobiographical memories *My World View*, Einstein mentions that those early thoughts on light already formed the basis of the special theory of relativity, which he formulated in 1905 while employed as a patent office clerk in Bern. This is an interesting detail because relativity is often associated with advanced mathematics, which Einstein certainly had not mastered at that youthful age – even if he had, contrary to legend, excellent grades as a student. Moreover, it later became apparent that Einstein's mathematical skills, though ranking on a high professional level, could not compete with those of David Hilbert or Élie Cartan.

And it was his friend Marcel Grossmann who taught Einstein differential geometry, a difficult field of math he used for his final formulation of general relativity. Einstein's intuition, on the other hand, was unmatched when it came to discovering fundamental, abstract principles that constitute the essence of a law of nature. He often made use of thought experiments, which means they could not be carried out due to technical difficulties (some were realized decades later), but are possible as a matter of principle. Many of these experiments became legendary and led to decisive insights.

The years of anxious searching in the dark, with their intense longing, their alternations of confidence and exhaustion and the final emergence into the light - only those who have experienced it can understand it.
- Albert Einstein

The special theory of relativity contained in the 1905 essay *On the Electrodynamics of* Moving *Bodies* dealt a heavy blow to the perfect world of classical physics. Space and time, since Newton the basic concepts of all theoretical physics, suddenly appeared to be strangely interwoven. It has been experimentally confirmed beyond doubt that moving clocks run slower, a phenomenon known as time dilation; even moving yardsticks become shorter. Another consequence of the theory was the formula $E=mc^2$, which nobody at that time suspected would predict nuclear explosions. However, the key idea on which all of Einstein's 1905 work was based had nothing to do with calculations: He realized that, according to Maxwell's equations, electromagnetic waves must propagate in empty space without any reference to the emitter. Therefore, since light was such a wave, its speed had to be independent of the motion of such a source or observer. All the rest follows from this single principle.

"I don't have any special talent, I'm just passionately curious."
- Albert Einstein

In a similar manner, the general theory of relativity, completed by Einstein ten years later, was based on a simple thought experiment, the so-called equivalence principle. It states that one cannot distinguish between the following two cases: 1) being on Earth in a dark room and perceiving one's weight or 2) being pressed to the floor with the same force because one is in an accelerated system in weightless space, for example in a rocket during propulsion. In the latter case, inertia of matter against acceleration is responsible, but in the former, it is gravity. Accordingly, this principle is also called *equality of inertial and gravitational mass.*

STUBBORN NATURAL PHILOSOPHERS

It is astonishing that modern physics, despite all its worshiping of Einstein the genius, is so little interested in *how* he actually achieved his success. Today, tens of thousands of physicists deal excessively with calculations that Einstein would have been unable to perform; yet for more than a century, no basic principle has been developed that can even faintly match the importance of Einstein's equivalence principle.

Einstein did not seek anyone's company (including that of his wife) when working. Sometimes, he insisted on his ideas in a thoroughly self-confident and even stubborn manner. Yet there is nothing conceited in his own words:

> I am not suited for tandem or team work .... Such isolation is sometimes bitter ... but I feel compensated for it, as it enables me to be independent of the customs, opinions, and prejudices of others, and do not seek to rest the peace of my mind on such shaky foundations.[46]

Special mention should be made here of the Viennese physicist and philosopher Ernst Mach, an important pioneer of general relativity whom Einstein held in high esteem. Mach, with his sound logic, is the prototype of the

European natural philosopher,[I] who was always engaged in a headlong search for the ultimate laws of nature without doing much calculation. In 1883, in his most important book *The Science of Mechanics: A Critical and Historical Account of Its Development*, Mach pointed out a weakness in Newton's theory of gravitation that physicists had been unable to spot for almost 200 years.[47]

A DEEP THINKER TO CHALLENGE NEWTON

> It is utterly beyond our power to measure the changes of things by time. Quite the contrary, time is an abstraction, at which we arrive by means of the change of things.[48]
>
> - *Ernst Mach*

Mach attacked the concept of absolute space and time introduced and allegedly proved by Newton with a thought experiment. Newton imagined a bucket filled with water.[49] As long as the bucket is at rest, the water surface naturally remains flat. Once one rotates the bucket around a vertical axis, friction gradually transfers the rotation of the wall to the water. As soon as the water itself rotates, its surface slowly climbs up the bucket walls due to the centrifugal force, eventually forming a curved surface. This curvature, on the other hand, proves that the bucket rotates, even if it seems to be isolated from the rest of the world. Newton

---

[I] This is not contradicted by the fact that Mach, as a positivist, dissociated himself from speculative or "meta" physics. He even doubted the existence of atoms. Before the discovery of the wave nature of matter, the criticism of a naïve particle picture was quite justified.

concluded that this proves that an unaccelerated, absolute space exists that does not rotate, without caring about whether the surrounding universe and the distribution of masses therein might influence the phenomenon he described. Mach came up with the following profound objection:

> Nobody can tell how the experiment would turn out if the bucket walls became increasingly thicker and more massive ... eventually several miles thick. . .

Evidently, he was insinuating that the water might perhaps no longer rise up at the walls if the whole universe rotated around the bucket. The distant masses in the cosmos would then define what it *means* "to be at rest." The inertia of masses that resist acceleration, would then, he suggests, be linked to the entire universe. Wow!

By further assuming the equivalence of the effects of acceleration and gravity (thus anticipating Einstein's principle of equivalence), Mach developed the bold idea that strength of gravity depends on all other masses in the universe, a hypothesis now known as Mach's principle. While the underlying problem has not been solved to this day, Mach's principle lives in the shadow of today's monumental constructs of theoretical physics; most "modern" U.S. theorists would shruggingly tell you it is philosophical claptrap from *Old Europe.*

Tragically, Mach's unfinished symphony could not be linked to data at the time; the first cosmological observations that would have validated his idea were not available until around 1930, 15 years after his death. And indeed, observations suggest that gravity has its origin in all masses in the universe! The gravitational constant G, which quantifies the strength of this force, is evidently related to these data.[I] The reason I have addressed Ernst Mach's idea in such detail is the importance of such fundamental constants as G from a natural philosophical point of view. The gravitational constant G, like the speed of light c, are enigmatic messages of nature that physicists have not yet been able to decipher.

THE FEWER CONSTANTS, THE BETTER

Whenever physicists actually succeeded in explaining such a constant – a rare case – it usually resulted in a revolutionary advance. For example, the essence of Maxwell's electrodynamics can be summarized by the formula $\varepsilon_0\mu_0 = 1/c^2$, which was confirmed by Heinrich Hertz in 1886. This was proof that the speed of light c, obviously determined by the electric and magnetic constants ε_0 and μ_0, was nothing but the speed of electromagnetic waves. A sensational discovery, methodologically reflected by the fact that the above equation reduces the number of independent constants of nature by one – a simplification.

The study of natural constants, which plays a prominent role in the European tradition of physics, is an approach to fundamental physics that emphasizes the simplicity of basic laws.[50] On the other hand, unexplained discoveries usually generate new constants and thus add to a complication of theories. It is therefore obvious that every natural

---

[I] Approximately, $c^2/G = M_u/R_u$ holds, R_u and M_u indicating the visible size and mass of the universe. Already in 1925, Erwin Schrödinger had already published this hypothesis in an article. Like many non-English publications from that time, Schrödinger's paper is quite unknown today. Instead, the above coincidence is usually linked to a superficial idea we will discuss soon.

constant is a problem yet to be solved. Einstein, being convinced that the basic laws must be simple, expressed it this way:

> I cannot imagine a reasonable unified theory containing an explicit number which the whim of the Creator could just as easily have chosen differently.

We are indebted to Ilse Rosenthal-Schneider – a philosophy student who got to know Einstein in Berlin around 1920 – for the fact that these and other remarkable statements of Einstein are documented. She also had to flee Nazi Germany in 1938 and later settled in Australia, from where she started her postwar correspondence with Einstein.

As mentioned above, the cosmological observations suggest that Mach's principle would allow for a calculation of the gravitational constant G. The coincidence $\frac{c^2}{\mathrm{G}} \approx \frac{M_u}{R_u}$ was taken up in the 1980s in the U.S. by some researchers who knew nothing about either Einstein or Mach. Soon, there was much hype about a theoretical fashion called "cosmic inflation," which did little more than produce a series of arbitrary numbers, instead of seeking fundamental explanations. It is easy to extrapolate how Einstein would have commented on this.

One thing that distinguishes Einstein from today's theorists is his expertise in virtually all areas of theoretical physics. It was this overview that allowed him to kick off the second scientific revolution of the 20th century, namely quantum mechanics, although he is famous for having created special and general relativity in the first place.

PLANCK'S BABY THAT EINSTEIN BROUGHT TO LIFE

Groundbreaking work for the quantum revolution had been done by Max Planck, who was born in 1858 and became a professor in Kiel in northern Germany in 1885. Planck's contribution also sheds light on the coevolution of technology and fundamental research. Soon after Edison had invented the light bulb, people tried to improve its efficacy by measuring the so-called black-body radiation, which Planck tried to understand at the turn of the century. However, Planck, the theorist, was rather disinterested in light intensity, but wanted to derive a law that described the emission of radiation of any wavelength (visible light covers only the range of 400–800 nanometers) at a given temperature. His mathematical skills enabled him to guess the correct formula, and later justified it theoretically.[51] In his newly found expression, a strange quantity appeared which he modestly called "auxiliary quantity"[I] h. It later turned out to be perhaps the most important fundamental constant of physics, nowadays known as Planck's quantum of action.

> People complain that our generation has no philosophers. They are wrong. They now sit in another faculty. Their names are Max Planck and Albert Einstein. – *Adolf von Harnack*

Einstein realized that h must have something to do with light emission and tried to use the idea for the still missing explanation of the photoelectric effect. In this experiment, incident light rays kick out electrons from metals, but the

---

[I] h abbreviates the German „Hilfsgröße."

results seemed paradoxical: below a certain cutoff light frequency, the electrons remained in the metal, no matter how intense the irradiation was.

In a bold leap of imagination, Einstein assumed that energy in a light wave was for some unknown reason quantized to amounts of E=hf (f denoting the frequency). The data brilliantly confirmed this hypothesis, although Max Planck, of all people, remained skeptical of Einstein's interpretation for years to come. These quanta of energy contained in a light wave later baptized the theory of quantum mechanics. However, until today there is no a priori explanation for those mysterious quanta. The phenomenon remained an unsolved riddle of nature. In the European tradition of natural philosophy, one would raise the question whether physics without this constant h is even conceivable, and if not, why not. Although unexplained, the quantum h nevertheless continued its triumphal path throughout theoretical physics. Even in thermodynamics – which one might consider rather far removed from other fundamental disciplines – h plays a decisive role, as for example in the mixing paradox named after the abovementioned William Joshua Gibbs.

All the fifty years of conscious brooding
have brought me no closer to answer
the question `What are light quanta?'
- *Albert Einstein*

UNMATCHED INTUITION FROM COPENHAGEN

The Dane Niels Bohr (1885–1962) was also one of the greatest thinkers in physics of the early 20th century, and his strength was certainly intuition rather than computing. For example, Bohr was the first to recognize that radioactivity, discovered in 1896, originated from the atomic nucleus, not from the atomic shell consisting of electrons. Prior to anyone else, Bohr had understood[I] that the chemical properties of elements in the periodic table were determined by the number of protons in the nucleus. Today, these facts seem so self-evident that it is easy to forget what kind of intellectual achievement their discovery represents. Bohr's intuitive approach to physics surely helped him to pursue a suggestion of Japanese physicist Hantaro Nagaoka, which Wilhelm Weber had already formulated in the 19th century: The atom behaves like a little solar system! The idea that electrons orbited around the atomic nucleus like planets around the Sun was incredibly intriguing to the researchers of that time.

Despite the world-shaking consequences that research would subsequently lead to, keep in mind that at that time, atomic physics could be seen as an intellectual playground that held no promise of further application. All these researchers were driven by curiosity alone: they simply wanted to get to the bottom of things.

Bohr realized that Planck's quantum of action h had the unit of an angular momentum and, in a flash of genius, he speculated whether this angular momentum could amount to that of an electron orbiting the atomic nucleus, similar to a figure skater rotating with outstretched arms. This led to the spectacular success of the model of the atom named

---

[I] For example, 1 proton means hydrogen, 2 protons helium, 3 protons lithium, etc. The suggestion came from the Dutch lawyer Antonius van den Broek, who was allowed to publish his idea in the journal *Nature in* 1911 – something that is unthinkable in today's scientific climate (cf. Kumar 2008, p. 86f.).

after him, since Bohr was able to explain the findings arrived at in 1885 by Johann Jakob Balmer, a math teacher in Basel (Switzerland). Balmer is another physics wizard worth examining.

Balmer had studied the spectral lines of hydrogen, which were already known at that time: a series of wavelengths[I] that correspond to certain spectral colors. With incredible patience and after years of pondering over the numbers, he discovered[52] a simple formula that precisely described the colorful behavior of the atoms. For Balmer, this whole enterprise had no practical use whatsoever; apparently, he was driven solely by his desire to uncover a hidden mathematical structure in natural phenomena. However, his formula still contained a riddle[II] which was brilliantly resolved by Niels Bohr. In an analogous manner, Isaac Newton had justified Kepler's laws, which the latter had also guessed while pondering over numbers for years.

THE LONG ROAD TO THE ATOM

The analogy between the physics of the very smallest and that of the planets was thus almost perfect. Yes, it was true that Bohr's atomic model still needed a theoretical justification, which was to raise many new problems in subsequent years. Nevertheless, his model represented a peak of understanding in the search for the origin of matter, which had been started by Democritus 2,000 years before.

Of course, we are providing a bird's-eye perspective of the history of science and missing some events that deserve more attention. But the main purpose here is to understand

---

[I] The wavelength λ is related to the frequency f via the relation $c = f\lambda$.

[II] It contained an unexplained value, the so-called Rydberg constant R, which Bohr then was able to calculate.

what mindset and what natural philosophical convictions the founding fathers of physics had when they arrived at their insights. However, this is not to suggest that all these scientific breakthroughs were achieved exclusively through the theoretical reflections of European natural philosophers; the second pillar of physics – and no less important – remains experiment.

A major role, for example, was played by Ernest Rutherford, who was born in New Zealand in 1871. Rutherford was an incredibly energetic guy who roared rather than talked; when ascending stairs, he used to take three steps at once. In 1908, he established his legendary laboratory in Manchester, which literally became the center of experimental physics at the time. Rutherford contributed to almost all major results on radioactivity (including the basic classification into alpha, beta, and gamma radiation)[I] and he coined the term "half-life."[II] With his famous scattering experiments, he was the first to prove that the idea of atoms being little solar systems was not mere fantasy. With his focus on experiments, his hands-on approach, his relentless productivity, and his simultaneous reluctance to entertain theoretical speculation, Rutherford certainly represented an "American" approach to physics. There are many anecdotes about Rutherford, and his statement "Don't let me catch anyone talking about the universe in my department!" became famous.

[I] Alpha rays are helium nuclei, beta rays are fast electrons, and gamma rays are high-energy electromagnetic waves.

[II] The time in which one half of a substance decays. Mind however that after two half-lives, there is still a rest of one fourth, etc.

Progress in physics always emerged when skilled and creative experimenters collaborated with theorists pursuing fundamental questions. This kind of interaction flourished in Europe in the early 20th century, but the culture of theoretical reflection was never to take root in the U.S. The way in which Ernst Mach, Albert Einstein, or Max Planck pondered over the laws of nature remained alien to physicists in the New World.

Just as the law of causality immediately takes hold of the awakening soul of the child, making it tirelessly ask the question 'Why?' – so the researcher is accompanied by this law throughout his life. - *Max Planck*

PHYSICS LOSES ITS INNOCENCE

Shortly after Bohr had developed his atomic model in 1913, World War I broke out – an epic catastrophe for Europe. Not only did scientific communication came to a standstill, but many scientists fell into an inner conflict. Few adopted the stance of British mathematician Harold Hardy, who outspokenly told the authorities that he was "ready to go out and be shot at any time," but unwilling to "prostitute my brain for war purposes." Indeed, most scientists participated quite willingly in the war, some even with patriotic enthusiasm, and their fields of research were put to military use. The mathematician Courant invented wireless earth telegraphy, which became essential for front-line communication; others developed the very important acoustic localization of cannons. The chemist and 1918 Nobel laureate Fritz Haber[53] developed poison gas, the first chemical weapon. Lise Meitner, a revered pioneer of radioactivity, even congratulated him on this achievement.[54]

Max Planck, Conrad Röntgen, and Wilhelm Wien certainly did not enhance their reputations by signing the "manifesto of the 93," a proclamation of "patriotic" scientists that denied any German responsibility for the outbreak of war. Tragically, many talented scientists died senseless deaths in the war, such as nuclear physicist Henry Mosley, who fell in the Battle of the Dardanelles in 1915.

In addition to the outstanding mathematician David Hilbert from Königsberg, who called the world war a "folly", Albert Einstein was one of the few pacifist-minded scientists who opposed the collective insanity. The English astronomer Sir Arthur Eddington, son of Quaker parents and conscientious objector, contributed to post-war international rapprochement in a special way: in 1919, he undertook an expedition that confirmed the predictions of Einstein's general theory of relativity by determining the deflection of light rays at the Sun during a solar eclipse. Overall, however, the war had destroyed much scientific spirit. It left deep scars wherever researchers from different countries had worked together peacefully in the pre-war period.

CONCUSSIONS IN THE OLD WORLD

World War I had decisively weakened the European empires, and it became increasingly clear that America was about to become the new world power. At that time, the world view of physics seemed to be resting on solid ground. Despite radioactivity and black-body radiation, which contradicted classical ideas, the laws of theoretical physics seemed quite complete. Leading physicists in America such as Abraham Michelson claimed that the future was all about applying these findings. Indeed, what was known in physics sufficed to trigger a rush of inventions in the U.S.; anyone in the field of science and technology had enough to do, and many were busy with applications and patents. By contrast, more far-reaching, fundamental questions were met with little interest.

The following years, however, witnessed a complete overturn of established convictions in classical physics, mainly by relativity and quantum mechanics, which subsequently proved to be incompatible. The enormous conceptual difficulties that arose were dealt with practically in Europe only, and it was in Europe that the crisis spreading into the whole of physics started. The first sensational findings of quantum theory turned out to already contain the germ of that crisis.

Chapter 6
Erosion
Physics between fragmentation and disorientation

The 1920s were also a golden period for physics. Unlike in economics, however, the crisis that put an end to this period did not hit suddenly, but arrived tiptoeing. It had, by the way, little to do with America or its culture of thinking, simply because no fundamental theoretical physics was practiced there yet. It seems that physics simply became too difficult.

Despite all the satisfaction that Bohr's atomic model of the "little solar system" provided, it suffered from a serious shortcoming: it contradicted Maxwell's electrodynamics.[I] Maxwell's greatest success had been the prediction that oscillating charges would generate electromagnetic waves, the existence of which had been spectacularly proven by Heinrich Hertz in 1886. However, every oscillating charge or technically equivalent orbiting charge must necessarily radiate waves. Therefore, an electron orbiting the atomic nucleus would inevitably lose energy and finally crash into it, making the whole model absurd.

By the way, physicists have to date not identified any general formula that properly predicts the radiation of an

---

[I] Nobel laureate Max von Laue therefore strongly criticized Bohr's model: "It's all crap!" (Kumar 2008, p.107).

arbitrarily accelerated charge – one of the unsolved problems that is usually swept under the rug today.[55] Generally speaking, experimental physics from that time on produced more puzzles than theoreticians were able to solve. For example, the obtained wavelengths of the atom's spectral lines agreed largely, but not exactly, with the predictions of Bohr's model. The first idea was to attribute this to a kind of self-rotation of the electrons, akin to the planets in the solar system, which, in addition to their orbit around the sun, also rotate around themselves. Dutch physicists George Uhlenbeck and Samuel Goudsmit described this strange behavior in 1925, calling it "spin".

GREAT SUCCESSES AND EVEN GREATER MYSTERIES

Together with the Dutch experimentalist de Haas, Einstein had already discovered in an earlier experiment that spin produced a magnetic field twice as large as expected; this was strange enough. Yet in 1922, the physicists Otto Stern and Walter Gerlach devised an ingenious experiment to investigate the behavior of electrons in more detail. Their results were stunning: While every material object may rotate around an *arbitrary* axis in space, electrons seemed not to be allowed to do so. Their "rotation" axis was either exactly aligned with the magnetic field or exactly aligned with the opposite direction. There is no reason whatsoever for why elementary particles behave in such a way, and it remains unclear why nature endows particles with such a strange property called spin at all.[I] From a classical point of

---

[I] Today, the comparatively superficial question of why some particles have half-integer and some integer spin is being investigated. The underlying problem, however, is still unsolved.

view, there is no necessity for such a complication, but it might already be a hint that our classical concepts of space, time, and matter break down at the micro level of atomic scales. It is already clear that the very idea of individual particles must be abandoned. Whenever two atoms of the same kind meet, they can no longer be distinguished.[I]

> Phenomena of this kind made physicists despair of finding any consistent space–time picture of what goes on the atomic and subatomic scale. – *John Bell*

Significant progress was nevertheless made with regard to the problem of radiation; progress that should make its inventor famous. Born in 1901, Werner Heisenberg, like a whole generation of theorists, had studied[II] with Arnold Sommerfeld in Munich. In his early career, therefore, Heisenberg had the opportunity to discuss his ideas with the leading physicists of the time; he even had a personal conversation in Göttingen with the celebrity Niels Bohr. *How* these discussions were conducted, however, sheds light on the scientific culture in the successful phase of atomic physics in the 1920s. Usually, people worked for themselves, but there was an intense exchange between the leading physicists, mostly in face-to-face conversations. In his autobiographical notes *Der Teil und das Ganze (The Part and the Whole),* Heisenberg describes the spirit of optimism that

---

[I] Experimentally, this was directly shown by the so-called Bose-Einstein condensation (Nobel prize in 2001), but had long before been proven by thermodynamics.

[II] Another student of Sommerfeld was Vienna-born Wolfgang Pauli, a godson of Ernst Mach, whom Heisenberg befriended immediately.

was in the air at that time. Describing a hike in the mountains he had undertaken, Heisenberg stated that it could be taken as a metaphor for the state of atomic physics: just below the clouds, one could already see the sunlight shining through and could thus guess the direction leading to the summit.

The breakthrough came in 1925. While recovering from a hayfever attack on the island of Helgoland, Heisenberg succeeded in describing the atomic states by means of so-called matrices, rectangular number schemes to which unusual calculation rules apply.[I] Heisenberg had learned from the mathematician Max Born in Göttingen that these then rather exotic objects existed in the first place. The math made Heisenberg realize that had to abandon the notion that an electron in an atom had a definite orbit at all. This was shocking, yet he took comfort from the fact that these orbits were not observable. This argument seemed a little dodgy and earned him the criticism of Einstein, with whom he discussed his model shortly thereafter in Berlin. Heisenberg, however, cheekily retorted that Einstein, too, had defended his theory of relativity by arguing that one must content oneself with observable quantities when measuring time.[56]

THE WORLD UPSIDE DOWN – LIGHT PARTICLES AND MATTER WAVES

Einstein expressed his appreciation for Heisenberg's idea, stating that he had "laid a big quantum egg," but remained skeptical. Not only was Heisenberg's approach highly mathematical, but it also did not seem very intuitive

---

[I] Thus, the usual law of algebraic commutation, e.g. $3 \cdot 5 = 5 \cdot 3$, does not apply.

to Einstein. In this respect, however, help soon came from a young French aristocrat named Louis Victor de Broglie, who had also drawn a bold analogy to one of Einstein's ideas. With his hypothesis of light quanta, Einstein had attributed properties of a particle to a wave phenomenon (light), and de Broglie therefore speculated whether particles (electrons) might also have a wave nature. This would at least be a promising idea for Heisenberg's problem of missing particle trajectories. Although de Broglie himself was not really convinced of his hypothesis of matter waves,[57] the idea was soon brilliantly confirmed by diffraction experiments with electrons in crystals.[58] It is significant that de Broglie had a big picture of unification in mind: his first step was to equate Einstein's expressions for light quanta, $E=hf$, and relativistic mass, $E=mc^2$ – perhaps the two most famous formulas in physics – now succinctly put together as $mc^2 =hf$. Surprisingly, this approach, printed in the first page of de Broglie's doctoral thesis,[59] is largely unknown today.

De Broglie's bold assumption of matter waves was taken very seriously by the Viennese physicist Erwin Schrödinger, who held a professorship in Zurich. Without further ado, Schrödinger derived a wave equation named after him, which also justified Bohr's atomic model, but was much more intuitive than Heisenberg's description.[I] Who did Schrödinger take his inspiration from? This is one of the amusing unsolved riddles of quantum mechanics, since he

---

[I] The physicists in Göttingen could have discovered Schrödinger's wave equation six months earlier if they had listened to a hint from David Hilbert. The latter had immediately associated the matrix calculus with differential equations.

had found his equation in 1925 during a ski vacation with an unknown lover.

A heated debate subsequently arose about which formulation of quantum theory was the "right" one. Heisenberg did not cut the finest figure when he overwhelmed Schrödinger with questions after the latter had given a lecture in Munich. However, the dust of the dispute settled somewhat, when Schrödinger[60] showed that the two versions were indeed equivalent. This was confirmed in 1926 by the English physicist and legendary introvert Paul Dirac, who had developed yet another approach to quantum mechanics.

EVERYTHING YOU THOUGHT YOU KNEW BECOMES BLURRED

Meanwhile, the hydraulic brake, the earth inductor compass, the liquid propellant rocket, and many other useful things were invented in the U.S. The extent to which the groundbreaking developments in atomic physics passed America by is remarkable, despite considerable funds being invested in the foundation of physics institutes, which no doubt carried out top-level research. For example, Andrew Millikan from the University of Chicago received the Nobel Prize in 1923 for his measurement of the elementary charge[I]. Arthur Compton, born in Ohio, received the same Nobel award in 1927 for the discovery of the effect named after him, which also proved the particle nature of light. However, with respect to the big questions of natural philosophy that troubled theorists, the U.S. remained a devel-

[I] Which also turned out to have a quantized value.

oping country. In 1929, Paul Dirac declined Edward Condon's invitation to Chicago with the offhand comment,[61] "There are no theoretical physicists in America." On the other hand, Dirac apparently found partners matching his level when he exchanged ideas with Russian theorists such as Tamm, Frenkel, Landau, Iwaneko, and Fock, whom he met on a journey to Leningrad and Constantinople.[62]

In 1927, Heisenberg succeeded in a catchy formulation of his key idea that would become famous as the "uncertainty principle". According to this principle, it is impossible for fundamental reasons to measure the position and velocity[I] of microscopic objects such as electrons at the same time. Intriguingly, this also holds true for other pairs of concepts such as energy and time, adding the appeal of an even more general theorem. The uncertainty relation can be observed in a multitude of physical experiments, yet it remains incompatible with classical ideas. Einstein strove hard to disprove the idea, but with no success; thus presumably contributing to the fame of the uncertainty principle.

QUANTUM CONFLICTS

Subsequently, theorists formed different camps, each of them cherishing one or another interpretation of quantum mechanics. Einstein was sympathetic to the approach of Schrödinger and de Broglie, while Heisenberg's formalism was supported by Niels Bohr, Wolfgang Pauli, and Pascual Jordan, as well as by the Göttingen mathematician Max Born, who introduced what today is known as "statistical

---

I More precisely, linear momentum, the product of mass and velocity.

interpretation" of Schrödinger's wave function: unlike other wave equations, it should not indicate a conventional density of matter, but merely a *probability* of encountering an electron at a certain place at a certain time.

This led to the inevitable consequence that electrons would decide by chance when to jump onto a lower atomic shell by emitting light. Schrödinger strongly opposed the idea that electrons would spontaneously perform such "quantum leaps" in a random manner; in his view, this contradicted Leibniz's credo *natura non fecit saltus*.[I] Wolfgang Pauli, in turn, called Schrödinger's lack of acceptance of Born's statistical interpretation "childish."[63]

Most people don't even see what kind of a dangerous game they play with reality. – *Erwin Schrödinger*

The widely known thought experiment *Schrödinger's cat* also helped to illustrate the conceptual absurdities that arise when one assumes that reality is a "superposition" of possible states governed by randomness. To make abstract reasoning graspable, Schrödinger imagined a cat in a box and vial of poison that could kill the cat in the event the vial were to break open due to a microscopic random event.

By analogy with the statistical interpretation of quantum theory, the cat would then be simultaneously in a "superposition" of two states, dead or alive, until someone opened the box to see what was wrong with the poor animal. Schrödinger was particularly allergic to the idea that reality would thus depend on the observer.

[I] (Nature does not make jumps). Allegedly, Schrödinger was quite colloquial when complaining: "The Göttingen folks are now using my beautiful wave mechanics to compute their shitty matrix elements."

Only a fool can deny the existence of a real external world.
- *Erwin Schrödinger*

Schrödinger received support from Einstein, who thoroughly objected to such a prominent role of randomness in the laws of nature, summarizing the critique with his legendary statement "God does not play dice!" However, Einstein soon specified that he had nothing against randomness in principle, but that such a law of nature should be derived somehow, not simply postulated.[I] Incidentally, Einstein by no means believed in an omniscient or even omnipotent God, but described his faith in terms of a deep, simple order that is behind the laws of nature as "cosmic religiosity".[64]

SHOWDOWN OF THE LUMINARIES

In 1927, the huge progress made in the field of atomic physics, as well as the unsolved contradictions of the new theories were considered a suitable topic to be discussed in a large conference of leading physicists. It was hoped that, in a joint effort, all these bright minds could eventually cut the Gordian knot of difficulties that seemed to have entwined the true laws of nature. The meeting, organized by the Vienna-born Paul Ehrenfest and financed by the Belgian industrialist Ernest Solvay, was to become the most consequential conference in physics ever. If there is any one moment that could be called the crisis of European theoretical physics, it was this meeting of 1927. Never again did a similar elite of physicists gather in one place, and never

---

[I] Following Cato the Elder, he used the phrase "probabilitatem esse deducendam," not "esse delendam!"

again were there such great expectations of reaching a breakthrough by such a unique exchange of ideas.

But collective endeavors, however bold they may be, rarely work in science. The development of quantum theory had never been never a team effort, not even in its revolutionary phase around 1920. All leading players struggled with their own approach, which they defended against competing ideas; rather than striving for a common vision, most preferred to work in intellectual seclusion.[65] The scientists of the time were truth seekers, and the idea of compromise had always been alien to truth.

As a consequence, some of the discussions at the conference resulted in chaos, while the intellectual mentor of the conference, Hendrik Antoon Lorentz, was overchallenged with the task of translating the discussions into the three languages: English, German, and French. The highlight of the conference was probably the debates between Einstein and Bohr. Einstein came up with numerous counter-arguments against randomness and the uncertainty principle, usually in the form of thought experiments that cost Bohr sleepless nights. Often, however, Bohr was able to refute the argument, once in spectacular manner by referring to Einstein's own general theory of relativity.[66]

. . . today's crisis in fundamental science highlights the need to revise its foundations down to an elementary level.[67] – *Erwin Schrödinger*

THE POISON OF RIVALRY

In terms of content, the "Copenhageners" Bohr, Heisenberg, Born, and Pauli argued that the formalism they had developed was a satisfactory description of the atom. Accordingly, they interpreted Schrödinger's wave function as a probability density, turning it into a tool that described the inherent randomness woven into the laws of nature. The mostly silent Paul Dirac agreed at best wordlessly[68] , while Schrödinger tried to recover a conventional interpretation of his equations. Like Einstein, he remained skeptical of the "Copenhagen interpretation" throughout his life. The inconvenient truth is that probably none of the versions offered is completely satisfactory. The persisting riddles of quantum mechanics actually suggest that we still do not properly understand the fundamentals of physical reality, space, and time.

> Many find it hard to admit that we are still in the toddler stage, and it is unsurprising that these guys do not want to concede that (myself included).
> *- Albert Einstein*

The respective merits of the theories presented at the Solvay conference can be debated endlessly. However, perhaps worse than the failure of agreeing upon a satisfactory theory were the consequences for the *kind and style* of the physics that was subsequently pursued. The main culprit of this unhealthy development was the ambitious Heisenberg. Undisturbed by self-doubt, he had declared himself the winner of the debate, which was not only factually incorrect, but promoted the fallacious idea that scientific truth could be established by a majority vote. For the first time, science became a little like politics, and theories were called "prevailing" or "established," as if this meant anything. This also shows that some of the deficiencies of contemporary scientific culture were not invented in America, even if they were later to become particularly common there.

The voice of the majority is no proof of justice. – *Friedrich Schiller*

Heisenberg, with his pursuit of success, was a character who represented this new way of thinking quite well. Always tackling problems in an optimistic mood, he glossed over the numerous open questions Einstein never became tired of emphasizing. In fact, natural science has always benefited from skeptics rather than from cheerleaders. Schrödinger, who also disdained the "method" of sweeping problems under the rug, wrote in his memoirs: "Instead of filling a gap by guesswork, genuine science prefers to put up with it." And he added venomously: "Once the problem has been eliminated by an excuse, there is no longer any need to reflect upon it."[69]

GENIUSES SCATTERED TO THE WINDS

After the conference, most of the leading figures went their own ways. While Heisenberg, at his new chair in Leipzig, continued to proclaim what he believed to be the solution of the problem, Dirac worked on a relativistic version of quantum theory, for which he was awarded a Nobel Prize in 1933, even if he later came to doubt his theory.[70] Schrödinger indulged himself in his affairs and later, turned his attention to general relativity. Pauli, after his mother's suicide, fell into depression and started drinking. De Broglie was a rather frustrated by the conference and continued to work in isolation; as did Einstein, who intensified his search for a unified theory of gravitation and electricity. Niels Bohr, on the other hand, buried himself in numerous treatises on the "complementarity" of waves and particles. Understandably, this was seen by many as a meaningless juggling of concepts and contributed significantly to the bad reputation that philosophers would subsequently enjoy among physicists. From this time on, "philosophy" became a pejorative term for theories that were thought to be mere gibberish.

The crisis of physics that emerged at that time originated in the fact that the true laws of nature were probably too difficult to reveal. Most likely, we are laboring under a fundamental misconception in our attempts to describe reality.[I] One of the most important unsolved riddles is the incomprehensible properties of spin. Given the conventional picture of space and time, there is no reason whatsoever for its very existence. Similarly, the existence of the natural constant h, whose units are linked to spin, cannot be deduced from fundamental principles.

> The question remains as to why Nature should have chosen this particular model for the electron instead of being satisfied with the point-charge. – *Paul Dirac*

Why is it impossible to design the laws of physics without that mysterious constant h? We do not know. These constants of nature, in a certain way, have taken over the role of modern gods – hitherto unexplained by the elite of physicists, they have been readily declared to be part of the unfathomable. Moreover, from an epistemological point of view, it is highly unsatisfactory that the world's future should depend on random events at the microscopic level.

[I] I argue that these difficulties are due to the fact that space and time are not the appropriate categories for describing reality, for which one can find good methodological arguments. For more details, see my book *The Mathematical Reality* (2019).

Though hardly anyone doubts the phenomenon of randomness, there is no deeper reason for it, as Einstein correctly remarked.

THE END OF DETERMINISM

French philosopher René Descartes was convinced that, in theory, the future could be calculated precisely as soon as the actual state of the world was known with sufficient precision.[I] This may sound naïve, and modern science has completely abandoned such a deterministic world view claiming that "nature is just (not) like that". Why? All other mysteries aside, the micro-world still holds a spin-related but independent problem that Einstein formulated in full clarity only in 1935; it became known as the Einstein-Podolsky-Rosen (EPR) paradox.

Everything is simpler than we can imagine, and at the same time more complex and intertwined than can be comprehended.
- *Johann Wolfgang von Goethe*

According to quantum theory, so-called entangled systems exist; for example, in the case of two electrons in the same orbital of a hydrogen atom, their spins always point in opposite directions, but if one looks at them individually, they are oriented at random. Thus, if one measures one spin, the orientation of the other is determined at the same instant. Since one can separate these electrons in space, an infinitely fast transfer of information to the other electron would have taken place as soon as one spin is measured.

[I] This is impossible even from a classical perspective, as research on dynamical systems, e.g. by Edward N. Lorentz in 1963 ('butterfly effect'), has shown.

This clearly contradicts special relativity. Nevertheless, this kind of magic long-distance effect has been experimentally proven.[71] This phenomenon is known as "nonlocality", a term that almost downplays the drama since it represents nothing less than a breakdown of elementary logic, given the conventional notion of space and time. None of today's physicists seems to have a big problem with that. The widespread belief that the Copenhagen interpretation leaves no questions unanswered is due more to post-war legend-making than to the objective consistency of the theory.[72]

Now we can start over; Einstein
showed that it doesn't work.
- Paul Dirac

These fundamental contradictions in our description of reality were soon joined by other problems. Beta decay – a form of radioactivity in which an electron is ejected from an atomic nucleus – appeared to violate the law of conservation of energy, one of the cornerstones of physics. Moreover, no one has given an explanation for the very existence of radioactivity to date. Theorists have not yet succeeded in deducing its origin from elementary principles. Unfortunately, one must therefore admit that the understanding of fundamental physics in Europe reached its limits in the late 1920s.

IN THE LAND OF LIMITED INSIGHTS

In the New World, people remained untroubled by all this – and correspondingly clueless. Despite all the hidden inconsistencies, the physics of the early 20th century offered so much technological potential that a wide field of activity opened up. The inventions from Europe were greeted with enthusiasm, and no one felt the slightest intellectual depression – unlike the European thinkers with their incomprehensible hunger for comprehension.

While in Europe the atomic theory [...] led to heated discussions [...] most American physicists seemed ready to accept the new view without any reservation.[73] – *Werner Heisenberg*

Of course, it is not the fault of Americans that nature is hard to understand; nor they are to blame for the failure of European physicists around 1930 to unveil its final secrets. But there was, and still is, a lack of awareness of these elementary issues in America. It simply did not occurred to anyone to address the logical contradictions of the Copenhagen interpretation of quantum mechanics. While people in Europe were racking their brains over the wave and particle nature of matter, the Americans simply called the puzzling phenomenon "wavicle," period. [74]

It isn't that they can't see the solution. It is that they can't see the problem.
- *G. K. Chesterton*

Fundamental physics was approached as an engineer would treat a bridge structure that could be tinkered with until it was sound. Today, such an approach is often called "effective theory," which is more or less a synonym for handicraft without understanding. Categories such as "final cause," "true reason," and "real understanding" did not exist among U.S. theorists – if anyone wanted to regard themselves as theorists at all. For all practical purposes, physics from Europe was more than good enough and contributed to the further rise of a nation that excelled in creativity and organization.

The visitor recognizes with amazement the supremacy of this country in terms of technology and organization.[75] - *Albert Einstein*

A FIRST GLANCE INTO THE COSMOS

One field of science in which the best instruments alone ensured a top position was astronomy, which until the end of the 19th century had played only a subordinate role in America. Since 1888, however, America possessed the most powerful telescopes in the world.[76] In 1906, a 61-inch reflecting telescope at Harvard College Observatory took images of unprecedented quality that enabled astronomer Henrietta Leavitt to identify an extra-galactic object for the first time. Leavitt had discovered that certain stars (Cepheids) in the Large Magellanic Cloud, at a distance of 170,000 light years, changed their brightness periodically. Her crucial discovery was that the smaller the stars were, the faster this happened, while slower changes were observed in bright stars. If they nevertheless looked faint from this great range, the distance could easily be calculated from the rhythm of their change. This ingenious method is still in use today.

Leavitt's discovery formed the basis for settling once and for all the great controversy over whether Andromeda was a galaxy proper or just a nebula in our Milky Way. In 1920, astronomers Harlow Shapley and Heber Curtis had conducted a public "Great Debate" on the subject, but it was not until the fall of 1923 that Edwin Hubble was able to identify a Cepheid star in Andromeda with his 2.5-meter instrument, proving beyond doubt its vast distance from

Earth. Immanuel Kant had been right in 1755 when he suggested the true nature of galaxies, which he called “island universes.” Hubble's famous Mount Wilson Observatory, built near Los Angeles in 1917, would remain the largest in the world for three decades.

PUZZLED BY RECEDING VELOCITIES

In 1929, Hubble found that the more a galaxy's light was shifted into the red, the farther away the galaxy was. Milton Humason was presumably involved in this discovery, too. He was a school dropout who had first worked as a porter at the observatory before scientists came to appreciate his talent. Since redshift was interpreted as recession velocity, Hubble's observation was considered evidence of an expanding universe.

If we dare to extrapolate back in time, this observation suggests a very dense initial stage, commonly referred to as the Big Bang. However, the original idea of such an early state of the universe goes back to the Belgian priest Georges Lemaître. Today, the Big Bang model is constantly being adapted to Einstein's theory of relativity through various additional assumptions, albeit with meager success, as I will explain later. In 1931, Hubble met his most famous visitor, Albert Einstein, in Pasadena. This caused a huge press frenzy, yet the scientific results were rather marginal.

Einstein then became involved in debates about various flavors of a cosmological model that rather obscured the connection to the origin of gravity identified by Ernst Mach. As mentioned in the previous chapter, Erwin Schrödinger's 1925 article[77] had already raised much more fundamental questions about cosmology than the models developed from the 1930s onward. This coincidence involving the strength of the gravitational constant, the speed of light, and the size and mass of the universe actually confirmed Ernst Mach's conjecture. But Mach, who died as early as 1916, seemed to be remembered by nobody at that time, not even by Einstein, whose mind was probably already more occupied with his unified field theory.

THE NUMBER OF THE UNIVERSE

Paul Dirac, who by then had worked for years to thoroughly understand the electron, was electrified by Hubble's observations,[78] because the data on the mass and expansion of the universe led him to a surprising connection. Dirac had long pondered why the electric force connecting the proton and the electron in the hydrogen atom was so much stronger than its gravitational counterpart. The ratio of the two forces is an incredibly large number of about 40 digits, the explanation of which Dirac had always considered a hopeless task until he came across Hubble's cosmological observations.[I]

To his utter amazement, Dirac found that the universe is larger than the proton by the same factor, which is already an almost unbelievable coincidence. Moreover, the universe is heavier than the proton by an 80-digit number (the square of the previous factor), which is additionally and independently a puzzling accident. These two observations are known as Dirac's large numbers hypotheses, which, ac-

---

[I] For the relation of forces $F_e/F_g \approx R_u/r_p$ holds, where r_p is the radius of the proton, R_u radius of the universe.

cording to their discoverer, point to a "deep connection between cosmology and atomic physics."[I] The association between the atomic nucleus and the cosmos is probably even more daring than the "small solar systems" of Wilhelm Weber, but no less fascinating. Despite the astounding agreement of the numbers, not only has nobody been able to explain this coincidence to this day, but it also seems to contradict all common models of modern cosmology. Tellingly, Dirac's hypotheses do not attract the interest of today's theorists,[II] despite the obvious validity of his argument.

I consider Dirac's thought to be one of the greatest insights of our epoch, the further study of which is one of our main tasks. - *Pascual Jordan, 1952*

THE CRISIS OF EUROPEAN THOUGHT

Step by step, I am getting used to the idea of not experiencing real progress.
- *Wolfgang Pauli*

A pattern similar to that of atomic physics in the late 1920s becomes apparent here. The difficulties of developing a satisfactory theory were obviously enormous, and the temptation to be beguiled by a plethora of new observations was strong. Because of the "late birth" of observational cosmology around 1920, it was hardly possible for a European tradition that sought explanations to develop. There are only certain scattered ideas in writings of people like Schrödinger and Dirac, while Ernst Mach did not live to see the verification of his profound ideas within a few years.

[I] The link to an idea of Einstein on variable speed of light from 1911 is discussed in my book *Einstein's Lost Key* (2015).

[II] An exception was George Gamov, born in Odessa in 1904, who even speculated about a connection to the fine structure constant.

The Europeans were aware of the crisis of their science. In 1932, on the occasion of a conference in Copenhagen, a parody[I] on Goethe's drama *Faust* was performed, in which Bohr took the role of God, Pauli the part of Mephistopheles, while Dirac was left to recite: "To 1926 we must return, our work since then is only fit to burn." Albeit in ironic form, the scene accurately reflects the time when the problems began.

This lack of success of the European tradition regarding fundamental questions later made it possible for physics to be taken over, so to speak, by American theorists; not because the latter had in any sense gained more insight than their European colleagues, but because they defined success differently. By hiding the problems that were too difficult, a sort of progress could be generated through observation and modeling that led, if not to monumental insights, at least to a solid self-confidence.

[I] The author was Max Delbrück, who first earned his doctorate in theoretical physics and later turned to biology, where he was awarded the Nobel Prize in 1969.

Chapter 7
Splitting Physics
Cheap explanations for unsolved problems

After quantum theory had already struggled with the problems mentioned in the last chapter, it suffered from even more in the late 1920s. In 1928, Paul Dirac formulated an equation that earned worldwide recognition because it combined Schrödinger's wave approach with Einstein's theory of relativity. This so-called Dirac equation led to interesting expressions that could be used to describe the *spin* of the electron.[I] Dirac's original goal, however, had been another: with his approach, he had hoped to calculate the mass of the electron, which was accurately measured soon after its discovery in 1897. In the European tradition, it was natural to try to derive an observed quantity. Actually, this is the core task of fundamental physics. Despite his intense efforts over a long period, Dirac did not succeed in achieving his great goal, and it became increasingly clear that there was a fundamental problem behind it.

In physics, there are three formulas that are so well established that no one would seriously doubt their validity: the distance law of the field of an electric charge discovered by Coulomb, another formula that quantifies the energy content of this electric field, and finally Einstein's famous relation between energy and mass, $E=mc^2$. However, if we

---

[I] However, contrary to a common claim, this does not constitute an *explanation of* spin.

assume a point electron, its field would contain an infinite amount of energy and therefore have infinite mass, which is of course nonsensical: the mass of the electron is undoubtedly only $9.11 \cdot 10^{-31}$ kg, a tiny but finite value. Antoon Hendrik Lorentz had racked his brain on this problem, the so-called self-energy of the electron. Conversely, all theories that assumed a finite expansion of the electron have failed.[I] Obviously, our conventional ideas of matter do not apply here.

A PRETEXT TO SAVE ENERGY CONSERVATION

Wolfgang Pauli was a fellow student of Werner Heisenberg in Munich, but they were very different characters. Heisenberg, the well-trained outdoorsman, loved hiking, while Pauli was fonder of nightlife and regularly slept through his lectures.[79] Heisenberg was generally an optimist, whereas Pauli's destructive criticism[II] could be cruel; however, because of his frankness he was also called the "conscience of physics." Referring to the above problem, Pauli flatly stated, "We will be regarded as the generation of physicists who left fundamental problems like the self-energy of the electron unsolved."

Pauli, however, had elsewhere damaged the culture that sought fundamental understanding. The results on beta decay (in which fast electrons are ejected from the atomic nucleus) seemed to indicate that some energy was lost in the

---

[I] Richard Feynman described the desperate efforts of physicists in his *Lectures* (vol. 2, ch. 28): "I am only showing you this so that you can see what crazy ideas people come up with when they know neither in nor out."

[II] Pauli's blasphemies sometimes even led to the person in question not publishing and thus missing out on the Nobel Prize, such as the Dutchman Ralph Kronig for electron spin.

process. Pauli first suspected that some kind of radiation had escaped the experimenters and mischievously expressed his doubts about their abilities, "I also think that they [...] are somehow cheating and have thus far overlooked the gamma rays due to their ineptitude."

But when the strange results were confirmed, he proposed, perhaps not entirely seriously,[I] a new particle (later called neutrino) that would carry exactly the missing energy. Against the background of the prevailing thinking, this was a cheap excuse, and he even reproached himself shortly thereafter: "Today I did something unpardonable. I replaced something that cannot be explained with something that cannot be tested." Paul Dirac also dryly remarked that, in his opinion, this was a trick invented for the sole purpose of "bringing the energy balance back into equilibrium."[80] Neither Pauli nor Dirac could probably have imagined at that time that neutrinos would one day become a research field of their own.

It is difficult to fathom what prompted Pauli to come up with his proposal. The missing clarification of the puzzles of quantum mechanics may have been playing a role. But perhaps, as he remarked ironically elsewhere, he also wanted to tease Niels Bohr and his working group, who were considering a violation of the law of energy conservation. Shortly before, right after his divorce, Pauli had started drinking. He later sought therapy with the well-known psychoanalyst C. G. Jung, which even led to a scientific exchange on psychology.

[I] This took place in 1930 in the form of a letter to the participants of a conference in Tübingen, whom he addressed as "Dear radioactive ladies and gentlemen." The letter ended with the excuse that a "dance taking place in Zurich" made his participation impossible.

THE PARTICLE ORIGINATING FROM DISPLACEMENT ACTIVITY

In physics, at any rate, Pauli was a kind of split personality, torn between the natural philosophical tradition of seeking to understand things properly and the pragmatic view of classifying unexpected results without too much reflection. What followed, the excessive complication of the neutrino model, was of course not Pauli's fault alone and would hardly have met with his approval. If he had not done so, then somebody else would probably have come up with the idea of postulating such a particle. But it is important to keep in mind that at that time no one regarded this as progress, let alone as a great discovery.[81] The neutrino explained nothing.

Pauli's two souls represented a schism that was soon to be felt throughout physics. On the one hand there were the difficult unsolved theoretical problems; on the other, the exciting, sometimes sensational, results of the experiments. This situation ultimately led to a kind of large-scale psychological displacement: instead of agonizing over theories that could conclusively depict reality but usually ended in frustration, physicists succumbed to the lure of the sweet and comparatively low-hanging fruit of observations that not uncommonly led to fame and Nobel prizes. The former, thoughtful approach apparently originated more from Europe, while America gravitated towards the latter, as the technology of experimentation advanced.

In 1932, Englishman James Chadwick discovered the neutron, an uncharged particle that spontaneously divides into a proton and an electron after about ten minutes – another manifestation of beta decay. Since then, for example, it is assumed that a helium nucleus consists of two protons and two neutrons, instead of four protons and two nuclear electrons, as was thought earlier. The former view was more thought-economical, however, as the neutron explained some nuclear properties better. There is not the slightest doubt about the reality of the decay, although the question remains why nature created such an unstable combination particle as the neutron in the first place. However, before anyone could ponder over this, the next surprise followed.

It has always been the dream of philosophers to construct all matter from a single elementary particle, so it is not entirely satisfactory that we have two in our theory.[82] - *Paul Dirac*

Then, at Caltech in Pasadena in 1932, Carl Anderson[83] discovered a positively charged electron, called a positron, which, along with a "normal" electron, disintegrates into pure light energy as soon as it hits its counterpart. This was first shown by English physicist Patrick Blackett in Manchester, who had also succeeded with the first nuclear transmutation in 1924. This phenomenon of pair annihilation, as well as the reverse process of pair creation of particles from pure energy, impressively confirms Einstein's formula $E=mc^2$, yet remains fundamentally mysterious. Why has nature even developed these two phenomena, light and matter, which can transform into each other in this fashion?

THE PLENTIFUL HARVEST OF TECHNICIANS

In the face of these indisputable results, theorists became less reluctant to postulate additional particles. Thus, the Japanese theorist Hideki Yukawa proposed a particle (later called pi-meson) that would allegedly mediate the nuclear force. A few years earlier, such a proposal would have been met with dissent, even ridicule. Nobody would have complicated the description of reality by further particles without compelling necessity. It would not have occurred to anyone to introduce another, third force merely to describe how the neutrons and protons of the nucleus are held together.

Nonetheless, Yukawa received wide acclaim for his proposal as early as the 1930s. It had a certain similarity with a theory developed by the Italian physicist Enrico Fermi on the description of the neutrino, shortly after this had been proposed by Pauli. Born in 1901, Fermi had received his doctorate from the University of Pisa in 1922 and went on to study with Arnold Sommerfeld in Munich. Competing against the many highly talented students there was certainly not easy, and Fermi gained the impression that his talent was not being sufficiently appreciated.[84]

But perhaps Sommerfeld was simply right. Natural philosophical thinking in the European tradition was obviously alien to Fermi, even if he later had tremendous success as an experimenter. In 1934, Rutherford, who was likewise more inclined towards practical work in the laboratory, congratulated Fermi on his "successful escape from the sphere of theoretical physics." Tellingly, Fermi later became regarded as one of the leading physicists of the postwar era in America.

It is interesting to consider the very different ways in which Enrico Fermi and Paul Dirac influenced physics. Fermi was an outstanding teacher. He playfully combined experiment and theory, but avoided the big philosophical questions; likewise, he was uninterested in the method, as long as the result fit. He was nicknamed the "quantum engineer"[85] for his practical approach, which helped to spread an engineering spirit among a new generation of American physicists.

THOUGHTS CIRCULATING IN THE IVORY TOWER

Dirac, who could rather be called an "ivory tower physicist," was completely different. He showed no interest either in teaching or in experiments, was taciturn, and preferred to work on his own. Dirac was captivated by the big questions and in pursuing them he was also concerned with methodological principles. His assertion that good theories must be "beautiful" is open to criticism, but mainly because this criterion can and has been abused easily.[86]

Simplicity, on the other hand, is certainly a criterion for the quality of a physical theory. Like no one else, Dirac saw the task of physics in theoretically calculating measuring values, in particular unexplained numbers such as the proton-electron mass ratio 1836.15. When Dirac searched for abstract mathematical structures, he did so to explain things that were truly relevant to physics. Fermi's efforts seemed naïve by comparison. Dirac would never have bothered to design a model of nuclear force by introducing a particle whose mass could not even be calculated. Dirac considered it explicitly nonsensical to turn to new particles before the masses of electron and proton were understood.[87] Nevertheless, Fermi's superficial way of modeling became widely adopted from 1940 on. Physicists like Dirac, on the other hand, seem to be extinct nowadays.

Whether by Yukawa or Fermi, much theoretical fiddling has been done in the meantime. But the truth is that to this day nobody really knows why we encounter atomic nuclei in their observed form and variety.[88] Assuming that there is a strong nuclear force with a short range does not exactly explain anything and already carries the poison of complication, which has never been a hallmark of good laws of nature. It should be borne in mind that Einstein never spent a single moment on the nuclear force, but for decades devoted all his energy to unifying the two fundamental interactions of gravity and electromagnetism. Apparently, he considered their strong interaction to be a meaningless concept that would become obsolete once the unification problem was solved.

> Indeed, from our previous experiences, we may trust that nature realizes the simplest thing that is mathematically conceivable.[89]
>
> - *Albert Einstein*

LONELY GENIUSES

One can only speculate what Einstein, Dirac, or Schrödinger would have thought about the fact that Fermi's description of beta decay – in its origin not understood to this day either – is today referred to as the fourth, so-called "weak" interaction. The arbitrariness contained in these concepts must raise the suspicion of every physicist in the natural philosophical tradition. Nevertheless, these four interactions are today regarded as taken-for-granted canon of theoretical physics.

When Einstein published his ideas about a unified field theory around 1930, which he had developed together with the French mathematician Élie Cartan, a completely different kind of physics was being practiced in the U.S.[90] Cosmic radiation, which constantly hits the Earth and in which Anderson had discovered the positron, became a popular playground for physicists. They were eager to detect other particles and ultimately identified a variety of exotic phenomena, such as the pi-meson – then interpreted as the "Yukawa particle" or the muon – which decays into an electron in a very short time. Nevertheless, upon the discovery of the muon, Isaac Rabi – a son of Galician immigrants who had in fact been assimilated into American physics culture – wryly asked, "Who ordered this?"

Why should Nature choose to make two particles that differ only in mass but are otherwise identical? All the above are examples of unexplained or unconnected facts.[91] - *Emilio Segrè*

Soon, these particles were also being generated and measured in the laboratory, which is not necessarily surprising at high energies, but fueled excitement about the apparent mass production of physical insights. The new measurements were made possible by the increasing number of new technologies. In 1930, for example, Ernest O. Lawrence, a son of Norwegian immigrants, had developed the cyclotron,[92] a device that could accelerate charged particles on circular paths to unprecedented velocities. In practical terms, it was a thumbnail version of today's largest particle accelerators.

AT THE FORK IN THE ROAD

For a proper understanding of the whole history of physics it is crucial to realize that these theoretical efforts in Europe and the observations being made mainly in America were two different worlds in a scientific sense as well. We are talking about two separate research paradigms that had nothing to do with each other in terms of content and therefore cannot be viewed as continuous progress. The psychological gap literally extended from one continent to the other; there was not even common ground for discussion. Einstein showed his utter contempt for the new particles in a polite way, namely with complete disinterest. But also among the European founding fathers of atomic theory like Bohr, Dirac or Schrödinger nobody was inclined to waste his time with theoretical models describing the wealth of newly discovered particles.

The same disinterest was shown by American physicists with respect to Einstein's attempts at a unified theory of electricity and gravitation, with respect to Mach's principle or Dirac's cosmological hypotheses – probably because these physicists did not really know much about these matters. American physicists participated in conferences with increasing frequency, as Arthur Compton did in the 1927 Solvay conference, for example; however, compared to the leading theorists, they rather played the role of craftsmen. It is not on record that any of them spoke up during, say, a discussion between Einstein and Bohr.

From these very different mindsets, correspondingly different notions developed of what progress meant. For Dirac, progress consisted in deriving theoretically unexplained numbers communicated by nature communicated; although he did not succeed in accomplishing this despite decades of effort. Even more puzzling than the above-mentioned proton–electron mass ratio 1836.15... is the so-called fine-structure constant $\alpha=1/137.035999$. It is the ratio of the speed of an electron in a hydrogen atom to the speed of light, surely an elementary property of nature. As a combination of different natural constants, the fine-structure constant can be measured precisely, but not theoretically derived.

As the legend goes, Dirac gruffly interrupted a young physicist who was about to present his new theory to him, asking, "Can you calculate the fine-structure constant? No? Come back when you have done it!" To this day, no one has ever succeeded in doing this calculation.

It's one of the greatest damn mysteries of physics: a magic number that comes to us with no understanding by man. [...] We know what kind of a dance to do experimentally to measure this number very accurately, but we don't know what kind of dance to do on the computer to make this number come out, without putting it in secretly! [...] All good theoretical physicists put this number up on their wall and worry about it.[93]
- Richard Feynman

A NAME THAT GLOSSES OVER THE PROBLEM

Interestingly, people talk about quantum electrodynamics (or "QED") as if a corresponding unification of these two theories existed. In fact, however, theorists in the 1930s tried in vain to reconcile the behavior of charges at high energies with quantum theory. Because of the blatant contradictions this involved, Dirac used to speak of "so-called quantum electrodynamics."[94] Nevertheless, legions of physicists work on "QED" today without caring about Dirac's objections. Actually, a theory worthy of this name should be able to calculate the fine-structure constant $\alpha = e^2 /(2 h c \varepsilon_0) \approx 1/137$, since both the "electric" constants and Planck's quantum of action h occur in it. This obvious failing resulted in considerable frustration among the European theorists, and only the most tenacious of them continued to work on it, usually on their own. Ultimately, however, the creators of quantum mechanics dispersed and capitulated in the face of the unsolved problems. Dirac's biographer, Helge Kragh, writes as follows:[95]

> Bohr, Dirac, Pauli, Heisenberg, Born, Oppenheimer, Peierls and Fock came to the conclusion, each in his own way, that the failure of quantum electrodynamics at high energies would require a revolutionary break with current theory.

However, this break was never to happen again. Instead, a new generation of physicists across the Atlantic – discussed in more detail in the following chapter – suppressed the problems and consoled themselves with other, thoroughly exciting things. Once again, Kragh aptly describes the situation:[96]

> With the recognition of the new particles (mesons) in the cosmic radiation, the existing theory – gradually improved in its details but not changed in its essence – proved to be quite workable after all. The empirical disagreements became less serious, and by the end of the thirties most young theorists had learned to live with the theory they adapted themselves to the new situation without caring too much about the theory's lack of consistency and conceptual clarity.

What else can be said about it? The European tradition of natural philosophy had given up on itself and would soon collapse altogether. This was certainly due to the tremendous difficulties that had arisen, but not exclusively so. For at the same time, the foundations of civilization, namely humanistic values, tolerance and international cooperation, were to fail in Europe on an even greater scale than physics.

FROM PEACEFUL COOPERATION...

With the founding of the German Empire in 1871 by Otto von Bismarck, who was very respectful to Jews,[97] they were protected from persecution. Although anti-Semitism was widespread throughout Europe, pogroms were considered uncivilized outbursts that had no place in the Prussian civil service state. Consequently, the Jewish community in Germany, especially in Berlin, flourished, even though many Jews had assimilated and the term therefore no longer really made sense. As David Nachmansohn points out in his book *German-Jewish Pioneers in Science*, researchers of Jewish descent played an important role in Germany's development into a leading scientific power. To name just a few of the most prominent, Max Born, Wolfgang Pauli, Otto Stern, Lise Meitner, Fritz Houtermans, Fritz Haber, Otto Meyerhof, Otto Warburg, and of course Albert Einstein come to mind. If we want to examine a particular biography more closely, we can take as an example the eventful life of James Franck, who in 1914, together with Gustav Hertz, proved the validity of Bohr's atomic model in an extraordinarily important experiment. Born in Hamburg in 1882 as the son of a Jewish banker, he developed liberal views as a child despite his religious upbringing. "Science is my God and nature my religion." he said of himself. He identified so strongly with the German Fatherland that he volunteered for military service in 1914, shortly after having carried out the crucial experiment for which he would later receive the Nobel Prize. After World War I, he got to know Niels Bohr in person and was appointed to a professorship in Göttingen, where he worked closely with Max Born.

... TO BARBARISM

I can only describe the current state of affairs in Germany as mental illness of the masses.[98] - *Albert Einstein*

Although Franck, as a war veteran, was exempt from the first Nazi racial laws, he left Germany in the fall of 1933, spent a year with Niels Bohr in Copenhagen, and eventually immigrated to the U.S., where he was offered a professorship at the University of Chicago. Later he worked on the Manhattan Project, with what feelings we do not know. Nevertheless, he had the almost unique courage to take a stand against the use of this devastating weapon and to point out the consequences with forthright words. Like others, he was ignored by the military.

Franck's students and colleagues appreciated his integrity, his human warmth, and his broad cultural background; he eventually died in Göttingen, which had made him an honorary citizen in 1953, after his first return to Germany. Even today, one cannot help but be enraged by the barbarism that forced people like Franck to leave their homeland. And yet he was only one among thousands[99] who suffered the same fate, while millions were killed.

This book can hardly add anything new to the elucidation of the roots of this disaster. The period was characterized by a fragmentation of society, intense ideology and emotionalizing of the public discourse, as well as an anti-intellectual atmosphere. The terms of the Treaty of Versailles, unwisely dictated by the Western powers and perceived as unfair by many, were certainly not a good breeding ground for peace. The Great Depression of 1929 caused economic hardship for many people, something that often resulted in a radicalization of the masses.

Nevertheless, having left the rule to the stupid, the history-less and the inhuman, remains a stigma on Germany. Even the intelligent opponents of the regime need to be reproached for having underestimated the danger. Einstein himself, during a visit to New York in 1930, airily remarked that the Nazis were a "childhood disease" that would soon be over. And Erwin Planck, Max Planck's eldest son and a high-ranking government official, told his Jewish friend Kurt Blumenfeld as late as December 1932 that he did not expect the Nazis to come to power; and that in any case, it would be impossible for them to hold on to power for more than a few months.[100]

Life is the highest value. The sacredness of the over-individual life entails the reverence of everything spiritual – a characteristic trait of the Jewish tradition.[I] - *Albert Einstein*

MADNESS DESTROYS SCIENCE

Once the Nazis had enforced their power by sheer force, shameful scenes took place in physics. The Nobel Prize winner Johannes Stark indulged in reflections on an "Aryan physics," disgustingly polemicized against Einstein, and did not even shy away from calling Werner Heisenberg a "mind Jew." Phillip Lenard, too, the discoverer of cathode rays, tarnished his own reputation by his remarks about

[I] This is very similar to the maxim of "reverence for life" of Einstein's contemporary Albert Schweitzer. Joseph Goebbels, the Nazi Minister of Propaganda, once wrote a letter to Schweitzer in which he wanted to win him over to Nazi ideology and concluded it "with a German greeting". Schweitzer politely declined, with "a Central African greeting."

Einstein. Einstein's correspondence with the German Academy of Sciences,[101] which ultimately led to his resignation, showed a deplorable self-denial on the part of those responsible. A few, like Max von Laue, had the courage to oppose the regime. Hitler's racism completed the already looming decline of European fundamental research, through the persecution of the most talented, the lack of intellectual freedom, and the focusing of science on weapons development.

Göttingen, one of the centers of top-level research, lost a particularly high number of scientists due to the expulsions, which deeply outraged David Hilbert. When asked in 1934 by the Reich's Secretary for Culture, Bernhard Rust, whether his institute had really suffered that much from the exodus of the Jews, Hilbert replied in thick dialect: “Suffered? It has not suffered, Sir. There is simply no institute any longer!”

Max Planck, with whom Einstein had a lifelong friendship, was also appalled by the banishment of Jewish scientists. He made use of his position as president of the Kaiser Wilhelm Society in Berlin to obtain an audience with Hitler. Planck courageously tried to convince the petty-minded dictator that the removal of Jewish scientists would only harm Germany, but received an angry rebuff.[102] Even more courageous was his son Erwin Planck, Max Planck's last surviving child.[I] He took part in the assassination plot against Hitler on July 20, 1944, which he paid for with his

---

[I] Planck's eldest son had been killed in World War I in 1916, and his two daughters both died of pulmonary embolisms shortly after the birth of their first child.

life in January 1945, despite a plea for mercy from his 86-year-old father.

Planck's personal tragedy, half a century after he had launched the most important development of modern physics, symbolized the ultimate destruction of European intellectual culture. For even among the victorious European allies, the loss of the scientific structures proved to be irreversible.

Part III
The Atomic Bomb and its Consequences

At the end of the war physicists
committed to war work had to consider
their future.[103]
- *Emilio Segrè*

Chapter 8
The Exodus of European Intelligentsia

America and the bomb

Despite its total defeat in 1945, Germany was spared devastation by atomic bombs. Since the discovery of radioactivity in 1896, this development was ultimately inevitable and, in some instances, could have happened more quickly. Only 10 weeks had passed between the German surrender and the detonation of the first nuclear bomb in July 1945.[104] Almost all fundamental principles of nuclear physics had been discovered by physicists from Europe or by Americans who had learned it there. As Robert Jungk notes in his book *Brighter than a Thousand Suns*,[105] nearly all the U.S. physicists involved in the construction of the atomic bomb had spent some time in Göttingen between 1924 and 1932. For a career in this field,[106] it was almost mandatory to have spent some time in Europe.

However, the most important nuclear physics laboratory in Cambridge, directed by the tireless Rutherford, lost its best collaborators before his death in 1937. Some of them immigrated to America, while others, like Pyotr Kapitza, were involuntarily repatriated to Stalin's Soviet Union. There, by the way, conditions for science were hardly better than in Nazi Germany: physics was valued, if at all, for the purposes of weapons development. In 1937, for example, the mathematician and theologian Pavel Florensky was sentenced to camp imprisonment and finally to death for spreading "counterrevolutionary propaganda". Guess what was meant by that: He had written a monograph on Einstein's theory of relativity. Russian nuclear physicists who later refused to participate in bomb development were arrested and deported.[107]

Nuclear physics, however, was a branch of science that relied on skillful experiments and original ideas in practice, not on deep theoretical understanding. In this respect, only Niels Bohr's insight that radioactivity originated from the atomic nucleus was crucial for development. After the discovery of radioactive radiation by Henri Becquerel, most things could be found out by careful and systematic trial and error.

JUST SHOOT AT EVERYTHING

Rutherford had already directed alpha rays at all sorts of substances, and beryllium proved to be the most interesting target. The bombardment released a neutron, which was not identified as such until 1932 by James Chadwick. Neutrons are the very key to nuclear technology because, as uncharged particles, they can approach any nucleus without a repulsive force and thus alter its composition. Some suspected that the neutron would one day release the enormous energies stored in the nucleus.

Hitler? Like all tyrants, he will break his neck in the not-too-distant future. Something else worries me much more deeply. Something that, if it falls into the wrong hands, will endanger the world more than this ephemeral fool. Something that – unlike him – we can never get rid of: the neutron.
- Paul Langevin,[108] 1935

Many laboratories were now busily working to direct neutrons at all sorts of materials. The leading centers were Paris with Frédéric and Irène Joliot-Curie, Berlin with Otto Hahn and Lise Meitner, while Enrico Fermi in Rome had recognized the efficiency of slow neutrons. But all these brilliant scientists did not realize over a period of almost five years that the heaviest element available, uranium, had long since been split by neutron bombardment.

God, out of His own unfathomable intentions, made everyone blind to the phenomenon of fission at that time.[109]
- *Emilio Segrè,* 1954

Instead, they believed that they had created heavier elements beyond uranium, the so-called transuranics, for which Fermi was even awarded the Nobel Prize in 1938. But almost all scientists were wrongly convinced that splitting the nucleus was only possible by much more energetic projectiles than neutrons. Only Ida Tacke-Noddack, a chemist in a comparatively low position at the University of Freiburg, had written in 1934 already that the uranium nucleus could disintegrate "into several large fragments." Hahn, despite her request, refused to even mention her hypothesis in a lecture in order "not to ridicule" Tacke. Even years later Hahn could hardly admit this arrogant mistake.[110]

THE DISCOVERER WHO DID NOT WANT TO LOOK

Otto Hahn's historical reputation as the "discoverer" of nuclear fission contrasts somewhat the obstinacy with which he failed to recognize the obvious, but nonetheless believed he was dealing with various isotopes and isomers[I] of the transuranics. Apparently, he was so confident of his extraordinary skills as a radiochemist that for a long time he did not pay attention to the results of the Paris laboratory; just like Lise Meitner, who had some scientific disputes with her great competitor Irène Joliot-Curie[111] that turned out to the advantage of the latter.

Eventually, in 1938, the Paris group published a paper that clearly indicated that a fragment of the uranium nucleus, lanthanum, had formed. But Hahn wouldn't even read it: "I don't care what the lady from Paris says!" Only when his assistant Fritz Strassmann literally forced him to at least take a look at the abstract did the scales fall from his eyes. Within a few days, Hahn had identified the fragment barium beyond doubt, thus providing the ultimate proof of nuclear fission.

Lise Meitner, born an Austrian citizen but of Jewish descent, had previously fled to Sweden with Hahn's help before the annexation[II] of Austria in March 1938 would have threatened her life. At the same time Hahn was writing the article, just before Christmas 1938, he wrote to her in Sweden asking her advice on how to interpret the results. Fortunately, Meitner was being visited by her nephew, the physicist Otto Frisch, with whom she immediately discussed the exciting news.

[I] Isotopes have the same number of protons but different numbers of neutrons. In contrast, the isomers of an isotope differ only in the energy state and the half-life.

[II] Presumably approved overwhelmingly by the Austrian population.

Apparently, due to the additional neutron, the uranium nucleus had become unstable and started to vibrate so intensely that it first contracted in the middle and finally broke into two parts. Meitner immediately developed this explanation as fission, which is still valid today, and sent it to the journal *Nature*. Now everything happened quickly. After his return from Gothenburg, Frisch met Niels Bohr in Copenhagen, who also suddenly realized that the nuclei had long since split.

How could we have overlooked this for so long? Oh what idiots we have all been, this is just as it must be!
- *Niels Bohr, 1938*

THE GENIE IS OUT OF THE BOTTLE

On his next visit to America, Bohr immediately blurted out the sensational news,[112] although he had actually promised Frisch that he would remain silent. The Hungarian physicist Leo Szilard, a brilliant mind far ahead of his time, was particularly alarmed. As early as 1932, he and his colleague Fritz Houtermans had been thinking about the enormous energy stored in atomic nuclei.[113] Among other things, they already realized that the almost inexhaustible amount of energy in the Sun could only be achieved by nuclear fusion of hydrogen to helium in its interior.

Szilard also anticipated the possibility of a nuclear chain reaction if additional neutrons were released during nuclear fission. Therefore, as early as around 1935, he tried to persuade physicists not to publish all the results of the perilous technology, but without success. In 1938, what had until then seemed to many a far-off speculation suddenly turned into a real danger. Using his own resources, Szilard conducted an experiment in Chicago in March 1939 that proved the feasibility of a chain reaction.[114] That night, as he wrote, he realized that "the world had started down a path of worry."

Together with his compatriot Edward Teller, Szilard ultimately prevailed on Einstein to sign a pre-written letter to President Roosevelt pointing out the possibility of a new type of bomb and the danger that Hitler might get his hands on it. As a result, a research project was launched, but the actual decision to proceed with the development of the bomb was taken later. Otto Frisch and Rudolf Peierls, who had both immigrated to England, probably contributed more than did Einstein's initiative. In their 1940 memorandum,[115] Frisch and Peierls for the first time provided specific calculations about the engineering design and the explosive power to be expected.

HITLER WITH THE BOMB - A NIGHTMARE

The motivating factor underlying the decision to develop the bomb was the fear that the novel weapon might already be in production in Germany, which for decades had been the center of atomic physics. Although many saw the possibility – including, of course, Heisenberg – the technical difficulties proved to be enormous. Again, it was Niels Bohr who first realized[116] that the fission reaction started from uranium 235, which occurs only as a tiny percentage in natural uranium. However, the separation of the isotopes 235 and 238 is highly complicated from a technical point of view and requires large-scale facilities such as gas centrifuges. In the later Manhattan Project, mass spectrometers[I]

[I] Invented by the British physicist Francis William Aston (1877-1945).

called calutrons were primarily used for this purpose, which had to “sort” some $3.6 \cdot 10^{-23}$ atomic nuclei individually for each gram of uranium 235. Heisenberg regarded this as a hopeless task, even more so under war conditions, and later wrote that the construction of a reactor he had begun served to produce energy, although in principle it was also capable of breeding[I] plutonium 239, which was equally suitable for weapons. However, there are reasons to doubt Heisenberg's account.[II] In any event, at the legendary meeting of the two in Copenhagen in September 1941, Bohr gained the impression that Heisenberg was indeed working on the development of the bomb and suspected that he was being sounded out,[117] while Heisenberg later claimed that his initiative had served to prevent a nuclear arms race.[118]

In any case, by this time several elaborate ideas for isotope separation already existed.[119] Of particularly explosive interest is that a May 1942 proposal on magnetic separation by physicist Heinz Ewald appears to be superior to the method later used at Los Alamos, as documents that have since been declassified show.[120] According to historical sources, physicist Manfred von Ardenne[121] took up this very clever method, having at his disposal a laboratory financed by the Post Ministry. Fritz Houtermans,[III] who had discov-

---

[I] Through the accumulation of free neutrons, uranium 238 first becomes uranium 239, which then transforms into plutonium 239.

[II] This is extensively documented in the book Hydrick (2016). Heisenberg's autobiography (1969) deals in particular with the period in question from the beginning of 1943 to the beginning of 1945 surprisingly briefly: "From the chaotic last years of the war, only individual pictures have remained clearly in my memory" (p. 222). It is true that Heisenberg was not particularly versed in technical matters (cf. Schirach 2012).

[III] Houtermans lived as a Communist in the Soviet Union from 1930-1937, but as a result of Stalin's purges he was arrested and tortured, losing all his teeth among other things. In 1938 he was handed over to the Gestapo, and

ered in 1940 that plutonium could also be used as fissionable material also worked in this laboratory,[122] which Hitler had allegedly visited several times. Further evidence suggests that the uranium was subsequently enriched on a large scale in a factory that ostensibly served the I.G. Farben corporation for Buna production.[123] More on this later.

In England, on the other hand, it was believed that not enough uranium 235 could be produced, partly because of the German bombings. Great Britain and the U.S. therefore decided in 1943 to coordinate their nuclear projects.

Incidentally, the final decision to produce the atomic bomb at great technical expense was made on December 6, 1941, on the eve of the Japanese attack on Pearl Harbor. Like other historical sources, this indicates that the U.S. had not only provoked Japan's entry into the war by imposing an oil embargo, but had also been specifically warned of the impending attack by decoding Japanese radio transmissions.[124] In any event, the public outrage over the Japanese attack allowed, without further inquiry, an immediate declaration of war by Congress on December 8, a pattern of American foreign policy that could be observed frequently in later years.

A LARGE-SCALE TECHNICAL EFFORT

The Manhattan Project, started for the development of the atomic bomb, was of gargantuan scale. Around the giant factory buildings, three cities were practically built from scratch: Los Alamos, Hanford, and Oak Ridge. Approximately 150,000 people were working on the project,

---

eventually gained freedom through Max von Laue's efforts.

without most of them knowing exactly what it was about, as strict secrecy was enforced.

Let us contemplate for a moment how the war and the destruction of scientific centers in Europe had transformed physics. Fifteen years earlier, the study of atomic physics was a mere academic pastime, motivated exclusively by curiosity to learn how nature was composed. Still, thinkers from many countries constituted a family that openly shared their results. Now, basically the entire intellectual elite were working in huge secret labs, preparing to seize world dominance with a new kind of weapon. Physics was never to recover properly from this disruption.

Throughout 1943 and part of 1944, our constant concern was that the Germans might complete the atomic bomb before we landed in Europe... In 1945, however, when we stopped worrying about what the Germans might do to us, we began to worry about what the United States government might do to other countries. [125] - *Leo Szilard*

All that was needed to build bombs was the application of existing knowledge. Technology was more critical than the laws of nature, and making things work was more important than understanding them. Teamwork took precedence over individual efforts, which in this case were even considered dangerous. At the time, the U.S. was probably the only country that had the economic and organizational capabilities to handle such a colossal task. Many energetic people were involved, among whom the military head of the project, Leslie R. Groves, was exceptional. Another outstanding figure was Hans Bethe, born in Strasbourg, who had studied under Arnold Sommerfeld and became director of the theory department. He, too, had been forced to give up his professorship in Tübingen in 1933 because of his partial Jewish ancestry. However, Bethe said he felt more at home in the dynamic and creative America than in Europe. Later, he was considered the leading theorist in the U.S.

Bethe had been appointed by Robert Oppenheimer, the head of the Manhattan Project. As the son of a textile merchant who had immigrated to New York in 1888, Oppenheimer had earned his doctorate in Göttingen a few years earlier, after Max Born had helped him overcome bureaucratic hurdles after a missed deadline. Dirac, on the other hand, who had shown little interest in nuclear physics experiments, refused to participate in the Manhattan Project. However, most of the leading physicists joined this endeavor – people such as Enrico Fermi, Rudolf Peierls, Otto Stern, James Franck, Eugene Wigner, Victor Weisskopf, Felix Bloch, Emilio Segrè, John von Neumann, and the aforementioned Hans Bethe; as did many young Americans who would later be heard from, such as Richard Feynman, Murray Gell-Mann, and Luis Walter Alvarez. Edward Teller, who because of the prevailing discrimination against Jews in Hungary had studied in Karlsruhe in Germany since 1920, also participated in the project after his emigration in 1933 and would eventually become known as the "father of the hydrogen bomb".

WAS GERMANY CLOSER TO THE BOMB THAN PREVIOUSLY ASSUMED?

Once thought, it cannot be taken back.
- Friedrich Dürrenmatt

As Heisenberg once remarked, in the summer of 1939 it would still have been possible to prevent the building of the bomb by a joint agreement of twelve people. But this was probably an illusion. Unnoticeably, the control of the project had slipped from the minds of the physicists into the hands of the military, who left no doubt that they would use the bomb in any event. Although there was a scientific advisory board, it was composed of people from whom not much dissent was to be feared: Oppenheimer, Fermi, Compton, and Lawrence.

The separation of uranium isotopes, which proceeded at a hopelessly slow pace at first, finally reached the "critical mass" necessary to ignite the bomb in the spring of 1945.[I] However, a relatively unknown event is perhaps historically significant here. Six days after the war ended, on May 14, 1945, the German submarine U 234, which had been straying in the Atlantic until then, finally surrendered and was brought into New Hampshire Harbor. According to existing documents, 560 kg of uranium, coated with a layer of gold, were unloaded in the operation.[126] Since gold absorbs neutrons, while "normal" uranium was available by the thousands of tons in the U.S. and Europe, this suggests that the shipment was indeed enriched uranium from the German Reich, originally intended for transport to Japan. Among the uncommonly high-profile passengers were two Japanese officers who committed suicide during the surrender. It is possible that that uranium accelerated the Manhattan Project by a few crucial months.[II] Whether this

---

[I] However, Hiroshima used about 50 kg of uranium, three times the critical mass.

[II] Other interesting details, such as the surprising installation of Admiral von Dönitz, the commander-in-chief of U234, as Hitler's successor, as well as the machinations of Nazi official Martin Bormann, who may have cooperated with the United States, can be found in the book *Critical Mass* by Heydrick (2016).

interesting alternative is actually correct will certainly continue to occupy historians.

In any case, the decisive "Trinity" test of the first bomb took place in the New Mexico desert on July 16, 1945, and its blast power exceeded all expectations.[I] Contrary to the wish of the majority of physicists, who had voted in favor of a demonstration over uninhabited territory,[127] the strike was being prepared over major Japanese cities. Two Navy officials even considered it unfair not to give any advance warning.[128] Moreover, the report initiated by James Franck only went as far as the advisory body, which did not oppose the will of the generals. After the test, which had caused an explosion more than ten times stronger than expected, Leo Szilard, in a final effort, drafted a petition against dropping the bomb without prior demonstration. Groves declared the paper classified so that it could not circulate among the scientists.[129] With that, the last attempt to save the lives of hundreds of thousands of civilians had failed. Groves was also unscrupulous in another respect: he prevented leaflets that warned of radioactivity from being dropped together with the bombs.[II]

[I] As in Nagasaki, the bomb was made from bred plutonium 239, while in Hiroshima enriched uranium 235 was used.

[II] Even their own population did not bother the military: the radioactive fallout was kept a secret, although the Kodak corporation had discovered it by chance on photographic plates. However, many people were not particularly squeamish. For example, Oppenheimer and Fermi proposed to poison the German population with the radioactive strontium-90 before the atomic bomb was ready (Schirach 2012, p. 129).

I told him I was against it on two counts. First, the Japanese were ready to surrender and it wasn't necessary to hit them with that awful thing. Second, I hated to see our country be the first to use such a weapon.[130]
- Dwight D. Eisenhower

MILITARY LOGIC TRUMPS HUMANITY

Although Japan was probably ready to surrender[131] in early August 1945, military rationale prevailed over humanitarian concerns.[132] The new type of bomb was used over Hiroshima and Nagasaki on August 6 and 9, 1945, resulting in more than 100,000 deaths from the explosions and countless more casualties from radiation poisoning. The German nuclear physicists came to know of this while imprisoned at the English country mansion in Farm Hall, Cambridgeshire. Otto Hahn was so shocked by the news that his colleagues were afraid he would commit suicide.

Leave me alone with your pangs of remorse! Look at what beautiful physics this is! *– Enrico Fermi, in early 1945*

In addition to the known consequences for the world as a whole, this event also transformed science permanently. Many of the greatest thinkers until the beginning of the 20th century were innocuous visionaries when it came to politics, whose work served only the noble goal of knowledge. The leading physicists from the atomic laboratories, on the other hand, were all of a sudden indispensable for the U.S. as a world power.

This had serious implications for the way science was organized. Researchers who wanted to return from Los Alamos, Hanford, and Oak Ridge to their civilian workplaces were subjected to secrecy regulations. Second, they were now also dependent on funding from the military, which often provided more than half of the research budget. When it came to the allocation of funds, Pentagon strategists were wise enough to leave scientists a great deal of freedom as to which projects they wanted to pursue. But most scientists first had to get used to dealing with problems of physics again on their own initiative. Many seemed to have no idea what to do with this new situation. In many fields, fundamental research was subsequently unable to shake off its proximity to the military. As a result, the hierarchies in research organizations also grew considerably.

> The scientists contemplating the state of theoretical physics descended into a distinct gloominess; in the aftermath of the bomb, their mood seemed postcoital.[133]
>
> - *James Gleick*

THE SCALABLE NUKE

The potential worldwide dominance through nuclear weaponry soon dictated the further course of events. Already in 1943 in Los Alamos, Edward Teller had insisted on working on the *Super*, the thermonuclear hydrogen bomb. Since the fusion of hydrogen into helium releases vastly more energy than nuclear fission, such a weapon has an explosive power of about a thousand times greater than a "normal" fission bomb. However, fusion itself can only be set in motion by detonating a fission bomb in the first place, which poses enormous technical difficulties.

While Teller, Álvarez, and Lawrence advocated the development of the *Super*, most researchers were initially averse to it. Even when, to the surprise of the experts, the first Russian atomic bomb test became public in August 1949, an advisory panel in October still opposed the development of the "super bomb". However, when in early 1950 the news broke that German spy Klaus Fuchs had leaked the atomic bomb program, including hydrogen bomb designs, to the Soviet Union, fear grew that Stalin would be the first to get his hands on the weapon.

Therefore, on January 31, 1950, President Truman announced that the U.S. would begin developing the weapon, though it was still not clear whether it would even be possible. Eventually, Teller and Polish mathematician Stanislaw Ulam came up with a viable design. The MANIAC computer, built by the Hungarian mathematician John von Neumann, contributed very significantly to its realization. Finally, on November 1, 1952, the first test of a thermonuclear explosion took place, wiping out an entire island in the Pacific.

THE TRANSFORMATION OF RESEARCH

Shortly afterwards, Soviet physicists around Andrei Sakharov had also invented a fusion bomb; in October 1961, the Soviet Union even ignited the most explosive "Tsar" bomb ever. The ensuing arms race of the Cold War is well known, and many high-hazard situations did not come to light until decades later.[134] With shorter advance warning times, the danger of nuclear Armageddon has certainly not diminished to this day.

The mindset of the future must make war impossible. - *Albert Einstein*

What is less in mind are the long-term consequences for science itself. Due to its applications, the field of nuclear physics was valued above all others. In addition, however, the kind of research that had led to this technological success began to dominate all of physics. The goal was no longer to understand the underlying processes, but to carry out large-scale projects that were believed to lead to new discoveries. On the one hand, this required constructing facilities using cutting-edge technology; but most importantly, it changed the position of the scientist himself. Instead of individuals, there were now huge teams organized in research collaborations working on a common goal. For example, whereas Michael Faraday, as a lone scientist, recorded about 10,000 experiments in his laboratory book, today 10,000 people are involved in *one* experiment at the European Organization for Nuclear Research (CERN).

Obviously, this is no longer the same kind of science. Among other things, the process of shaping opinions has changed greatly. In the end, the members of an institution must speak with one voice about how to interpret often complex experiments. Given the institutional hierarchies, the implications for *how* discussions would now take place were serious. It is not for nothing that sociology of science has become an important field in recent years, even if its findings are not always appreciated.[135]

> Science is seen as a process of implementation; the process is seen as something explicit and managed, and the role of the individual researcher is – in a nutshell – obedience.[136]
>
> \- *Bruce G. Charlton*

REQUIEM FOR PURE SCIENCE

In terms of its scientific content, the field of nuclear physics, which had such an "important" practical application, leaves many questions unanswered. The calculations on which bomb construction were based obviously did not involve anything fundamental. Moreover, even scientific nuclear physics is still far from being able to quantitatively predict the phenomena observed by theory. There is not a single nuclear transformation, not a single energy level that can be calculated from a quantitative theory. No one knows why the fission of uranium releases just such an amount of energy; no one can calculate the mass of the nuclear building blocks; no one can understand why they have just the size they have; no one can figure out why certain isotopes are stable and others are not; and certainly no one understands why radioactivity exists at all.

However, on July 16, 1945, all these questions became irrelevant. The science of Einstein, Mach, Schrödinger, and Dirac had ceased to exist. The new leaders from Los Alamos – Oppenheimer, Bethe, Lawrence, Compton, Fermi, Chadwick, as well as many young *war boys* who had learned their craft during the war – had assumed the intellectual scepter and subsequently defined what was regarded as important physics.

Chapter 9
After Hiroshima
New masters, new physics

Some physicists may be happy to have a set of working rules leading to results in agreement with observation. They may think that this is the goal of physics. But it is not enough. One wants to understand how Nature works.[137] - *Paul Dirac*

With the end of the World War II, theoretical physics had practically dissolved in Europe, but not the problems that had been known long before. The most serious of these was, as already mentioned, the infinitely huge self-energy of the electron, a completely nonsensical result, which however follows from the consequent application of the well-established laws of electrodynamics.

Actually, it should be obvious to everybody that this contradiction points to a fundamental shortcoming of these laws. Nevertheless, theoretical physics acts to this today as if the mishap could be remedied by a computational trick, without touching those formulas that lead to the wrong result. The term "renormalization" was coined for this procedure. I discuss this in more detail since it is relatively easy to understand, but also because theoretical physics has in the interim built many layers of ingenious constructions, which are based on this non-existent fundament. And although some physicists have rightly criticized parts of this theoretical construction, hardly anybody mentions the underlying absurdity of all this. The unresolved contradiction of self-energy and its pseudo solution of renormalization is a kind of original sin of modern theorizing.

My aim is: to teach you to pass from a
piece of disguised nonsense to
something that is patent nonsense.
- *Ludwig Wittgenstein*

To justify renormalization, it is argued that the electron has an infinitely large "naked" mass (whatever that is supposed to mean), from which one needs to subtract[I] the above-mentioned "electric" mass, which is also infinite, so that the difference results in – voila! – $9.11 \cdot 10^{-31}$ kg, the measuring value of the electron mass. This does not even become convincing if one puts it more precisely. For every mathematician knows that one must not mistake the term infinity (which is only justified as a limiting value) for a number. Otherwise one could deduce any nonsense from it. Nonetheless, theoretical physicists juggle with this mathematical trick to this day, and it is even regarded as progress if one finds further "renormalizable" theories.[II] As often in history, this shows that groupthink can also prevail over high individual intelligence and that the mechanisms of collective displacement are powerful.

The fact that an opinion has been widely
held is no evidence whatever that it is
not utterly absurd. - *Bertrand Russell.*

[I] Nowhere in physics has the concept of a negative mass ever made sense.

[II] Pauli and Dirac probably would not believe it, but in 2004 the Nobel Prize was indeed awarded for the invention of further "renormalizable" theories.

ABNORMALIZATION OF PHYSICS

In his 1945 Nobel Prize speech,[138] Wolfgang Pauli explicitly stated that one should not use "tricks" to subtract infinities. We have to be grateful that a reasonable physicist of the pre-war period has expressed himself at all on the subject. Einstein, Bohr, and Schrödinger probably considered the idea too outlandish to even give it a mention, but at least Dirac formulated a dry comment to which there is actually nothing to add:

> This is just not sensible mathematics. Sensible mathematics involves neglecting a quantity when it is small—not neglecting it just because it is infinitely great and you do not want it!

The only astonishing thing is that physicists continued to deal with renormalization at all. However, it is even more strange that this notion became the generally accepted canon of theoretical physics after the war. This was largely due to a conference held on Shelter Island in Massachusetts in the spring of 1947. The participating physicists were relieved that it took place at all. For after the war, the military had gained a foothold at all the universities and funded fundamental research.[139] Correspondingly, secrecy still reigned and loyalty commissions interfered with everything.

STARTUP OF A FAILED BUSINESS

For the first time, physicists who in previous years had been working mostly on radar technology or in the Manhattan Project could talk openly again at Shelter Island without security precautions. The meeting was a kind of reboot of fundamental theoretical physics in America. It contrasted strangely with the legendary Solvay Conference in 1927, which had revealed the fragmentation of European research. Twenty years later, the idea was to create a common basis for the theoretical physics of the future. However, this happened under completely different circumstances. At both conferences, the organizers sought to gather the elite, and in 1947 almost only American physicists were invited. Nothing can better illustrate the disruption of traditions than the fact that, apart from the Dutchman Kramers, not a single physicist attended *both* conferences. What had dissolved[I] in 1927 bore little resemblance to what reassembled in 1947.[II]

Experimenters William Lamb and Isaac Rabi reported their measurements for the first time at the Shelter Island conference. Lamb, who during the war had worked as an expert in microwave engineering, had used these to measure a small difference between the energy levels of the hydrogen atom (the Lamb shift), while Rabi had obtained a new value for the magnetic moment of the electron.

Inspired by that meeting, the aforementioned Hans Bethe calculated an estimate for the Lamb shift for the first time, while the young American theorist Julian Schwinger published a calculation of the magnetic moment of the electron. From the pre-war perspective, these activities of the new elite had a funny touch: instead of answering the old unsolved questions, they provided new, unasked-for solutions.

[I] Before the war, in 1930 and 1933 there were other Solvay conferences, but they did not match the importance of the previous one; cf. Jones (2010).

[II] At a subsequent meeting in Pocono in 1948, Dirac and Bohr attended, both of whom were very skeptical of Feynman's lecture, though this was probably due to misunderstandings.

It is striking how today's physics is capriciously focused on these two measuring results, which, at least according to the claim, can be precisely calculated; although they are by no means particularly salient properties of nature. We hear relatively little about the corresponding magnetic moments of neutron and proton, which seem hopelessly inexplicable with their outlandish numerical values.[I] And the much more important fine-structure constant $\alpha \approx 1/137$ is not computed by the theory anyway, although the self-assured name "quantum electrodynamics" should actually aspire to this.

CHOIR OF ENTHUSIASM

However, in the pragmatic view of the young American physicists, this did not matter much. Thanks to the new ideas, “renormalization” was increasingly seen as established; the majority of physicists thought that everything was fine and that the revolution they had longed for earlier was unnecessary.[140] Keep in mind that since 1950 virtually all attempts to describe the world of elementary particles were based on the absurdity of “renormalization”.

> Renormalization in QED is a drastic departure from logic. it changes the whole character of the theory, from logical deductions to a mere setting up of working rules.[141]
>
> \- *Paul Dirac*

If one addresses physicists on the conceptual problems of quantum electrodynamics, one is immediately told how

---

[I] 2.79 for the proton and -1.91 for the neutron. Otto Stern had already measured the former in 1933.

magnificently this theory has been tested by experiment. They claim that quantum electrodynamics predicts the magnetic moment of the electron up to the 12th decimal place,[142] which makes any further discussion superfluous for most people. This exemplifies how important an approach based on the history of science is for assessing theories. For very few physicists know the historical facts that thoroughly undermine the belief in the accuracy of quantum electrodynamics. These have recently been presented in detail by the Spanish researcher Oliver Consa.[143]

If one takes a look at Bethe's original paper, for example, one finds that he had inserted an arbitrary limit in his calculation which avoided the "infinities" and only for that reason led to the desired outcome.[144] After the experimental value had been optimized, Bethe corrected it in a later work, whose only justification was to give the desired result.

Julian Schwinger's celebrated first paper[145] on the subject consists of a single page in which he postulated, without any substantiation, a numerical relationship between the magnetic moment of the electron and the fine-structure constant – a detailed theory that Schwinger claimed at the time was in preparation.

PROOF BY FALLING ASLEEP

However, Schwinger did not present it until the next conference in Pocono in 1948. His presentation lasted no less than six (!) hours and was so tedious in its abstractness that only Bethe and Fermi remained in the room until the end – admittedly without understanding the calculations. The lecture by Richard Feynman did not receive more positive feedback either. He presented a scheme of diagrams with which he outlined a theory that allegedly reproduced the measured values exactly. However, when challenged, he was unable to offer any justification for his approach, especially any mathematically consistent calculations. In vain he tried to convince the audience with the argument that the accurate agreement alone proved the correctness of his thesis.

Others gave talks to show how to do the calculations, while Schwinger gave talks to show that only he could do it. – *Robert Oppenheimer*

While Schwinger had presented mathematics without any illustration, Feynman worked graphically, but without any mathematics. Curiously, both admitted not having understood anything of each other's approach,[146] but agreed to have each developed two versions of a theory which explained the observations in an excellent way.

The third to come on board of this strange enterprise was the Japanese physicist Isihito Tomonaga, who in 1949 sent a theoretical paper to Robert Oppenheimer, which also claimed to account for the experimental results, although it was hardly more transparent than that of Schwinger. The latter's original papers[147] contain 469 formulas, many of them over several lines, while in the first paragraph one is told that everything is just an approximation. So what should this really have to do with the simplest particle in the universe?

The electron is too simple a thing to ask about laws that give rise to its structure.
- *Paul Dirac*

It was now received with great relief that the English mathematician Freeman Dyson demonstrated the mathematical equivalence of these theories[148] – although Dyson had actually developed a version of his own. Dirac, Oppenheimer and Fermi scathingly criticized Dyson's work in personal conversations,[149] which apparently caused him to reflect; the reaction of the *community*, on the other hand, was enthusiastic: quantum electrodynamics was now considered established and, at the same time, already the state-of-the-art and most precise theory.

Dyson's approach assumes that the anomalous magnetic moment of the electron can be expressed as an infinite sum, with each term carrying a "weight". These weights – we might, less respectfully, call them *fudge factors* – can only be determined by highly complicated calculations. This type of model provided a gateway to what was to become a subtle self-deception that has continued to spiral in bizarre ways ever since. For the calculation of the "weights" is not grounded in legitimate mathematics, nor have its principles ever been spelled out clearly.

SUSPICIOUS SYMBIOSIS OF THEORY AND EXPERIMENT

An example of a downright scientific scandal was the calculation of one of the factors by two of Feynman's assistants, Karplus and Kroll, in 1950, who obtained a value of -2.973, which exactly matched a measurement published shortly before.[150] If one looks into their paper, one is surprised to learn that the decisive part of the calculation was not printed because the corresponding Feynman diagrams were "too complicated." Substantial parts of quantum electrodynamics, which today is considered "the best theory of physics," have to date still not been published![151]

As a detail, on top of that the result of Karplus and Kroll was wrong. After experiments again showed a different value, other researchers corrected the weight,[152] from -2.973 to -0.328, a weirdly large discrepancy. To this day it is not known what the error of Karplus and Kroll had been. But both conceded that their claim that the result had been checked by independent calculations had been untrue.[153]

> The pragmatic adjustment of experimental techniques according to their success in displaying phenomena of interest...[154] - *Andrew Pickering*

Even greater doubts about scientific integrity were raised by two experiments,[155] in 1961 and 1963, which exactly coincided with the respective theoretical value, and which formed the basis for the 1965 Nobel Prize awarded to Feynman, Schwinger, and Tomonaga. At this point, it may no longer surprise the reader that these experimental values turned out to be wrong and were subsequently adjusted in the direction of the latest theoretical value.[156] In fact, over time it has become increasingly difficult to identify the mutual tweaking of the theoretical and experimental results, which is evident from the papers published at that time. Meanwhile, the calculations[157] have become so complex that they can only be carried out by software, which, in practice, nobody can check. Also in the more recent past, theoreticians have made mistakes in counting the possibilities – something that had previously mostly gone unnoticed – so that even a wrong value for the fine-structure constant α was published.[158]

ROTTEN TO THE CORE

That was of course a terrible blow to all my hopes. I really meant that this whole program made no sense.[159] - *Freeman Dyson*

Besides these more-than-suspicious inconsistencies, there is another result that deprives quantum electrodynamics of any foundation. Freeman Dyson, perhaps cured by the criticism he had received, once again explored the premise of the entire theory and proved in 1952 that the series he had proposed does not converge[I] at all, in other words, for mathematical reasons it cannot yield any meaningful result in the first place![160] It is incredible that such a death sentence for a theory has been hushed up in a science that calls itself "exact." As a result, Dyson abandoned any involvement with quantum electrodynamics and, after returning to England, devoted himself to other fields. In 2006 he confessed in a letter:[161]

> I remember that we thought of QED in 1949 as a temporary and jerry-built structure, with mathematical inconsistencies and renormalized infinities swept under the rug. We did not expect it to last more than 10 years before some more solidly built theory would replace it. Now, 57 years have gone by and that ramshackle structure still stands…

Almost a little ashamed, even Feynman himself admitted the shortcoming:[162]

> Having to resort to such hocus-pocus has prevented us from proving that the theory of quantum electrodynamics is mathematically self-consistent. It's surprising that the theory still hasn't been proved self-consistent one way or the other by now; I suspect that renormalization is not mathematically legitimate.

[I] Convergence denotes the steady approximation to a limiting value, for example in the series 1/10+1/100+1/1000+... = 1/9.

All in all, Oliver Consa's recent research[163] on the history of quantum electrodynamics leaves little doubt that a decades-long combination of abstract wishful thinking, little dishonesties, confirmation bias, and groupthink has created a mirage of a theory whose evaporation in the light of history will still shake physics considerably. In any case, however, the ball is now in the court of the avid modelers, who must finally make transparent their assumptions and calculations.

PAINTED OVER RUST

It is obvious that the basic flaws of quantum electrodynamics were only glossed over with opaque methods, and the alleged "precision" is being parroted everywhere without anyone being able to check it. In 2021, spectacular measurements[164] were made on the magnetic moment of the muon, which – once again – did not agree with the theory. Nevertheless, quantum electrodynamics is still regarded as an untouchable truth, so for the time being people will probably fall back on other excuses instead of questioning it in its entirety.

> The work of Einstein, Bohr, Heisenberg, Schrödinger ... arose from deep thought on the most basic questions surrounding space, time and matter, and they saw what they did as part of a broader philosophical tradition, in which they were at home [...] in the approach to particle physics developed and taught by Feynman, Dyson, and others, reflection on fundamental problems had no place in research.[165] - *Lee Smolin*

If we look at the models developed since 1950 from a wider perspective, it becomes clear that these calculations with their types and logic have nothing to do with the successful physics of the beginning of the 20th century. The break with the earlier tradition is evident here. Feynman's approach elucidates the random nature of quantum mechanics, with the diagrams named after him, depicting all imaginable particle trajectories with the corresponding conversion processes. Nevertheless, in the end it is a playfully naïve approach, and it is clear that it does not truthfully describe the microscopic events. At the times of Einstein and even Lord Kelvin, who were always seeking for vivid-intuitive explanations, such abstract schemes would have been unthinkable. Everybody would have asked: What does this have to do with reality?

Quantum mechanics describes nature as absurd from the point of view of common sense. And yet it fully agrees with experiment. So I hope you can accept nature as She is - absurd.[166]
- *Richard Feynman*

MODERN THEORY BUILDING

One of the key characteristics of the paradigm shift from pre- to post-war physics is a kind of general acquiescence that physics is now permitted to look crazy. Demanding clarity, on the other hand, is frowned upon and seen as a lack of understanding of "modern" methods. The roots of this twisted perspective are actually to be found in the enormous difficulties that arose at the scale of atoms and nuclei in the 1920s. These problems should have been thoroughly understood, instead of accepting fanciful models that no longer aspire to depict reality. Clarity and vividness were replaced by computational rules, which were associated at a metaphorical level with the experiments. Many physicists will regard all this as a modern-day insight. But the standstill of the last decades rather suggests a path of degeneration.

With the loss of the visual instinct, physicists have lost a crucial tool for intuitively separating the wheat from the chaff when it comes to new ideas. Unfortunately, there is no longer anything that sounds too outlandish to be worth discussing. If then, predictions are so vague that an experimental falsification is difficult, then the survival of a veritable "model" is guaranteed.

The popularity of quantum electrodynamics and the related Feynman diagrams have led to a monoculture in the minds of physicists, who assume that all interactions can be described by an exchange of particles. However, there is not the slightest hint that this will lead to unification – apart from the fact that the classification into four basic forces is already unsatisfactory. Nevertheless, in the same scheme, "quantum field theories" such as "quantum chromodynamics" were later designed and praised as new insights. With his typical frankness, Feynman himself even made fun of it:

> So when some fool physicist gives a lecture at UCLA[1] in 1983 and says, "This is the way it works, and look how wonderfully similar the theories are," it's not because Nature is really similar; it's because the physicists have only been able to think of the same damn thing, over and over again.

---

[1] Abbreviation for University of California Los Angeles.

THE SUNNYBOY OF PHYSICS

Feynman's outspoken and humorous personality made him the most popular figure in physics. His memoirs *Surely You're Joking, Mr. Feynman!* are fun to read, not only for physicists. Also his book *QED: The Strange Theory of Light and Matter* became a bestseller because of its clarity. Feynman proved his straightforward character as late as 1986 when, in a quagmire of agency incompetence and attempted cover-ups, he uncovered the cause of the Challenger disaster.[167] In a short demonstration to the press, he showed that a rubber ring not resistant to cold had led to the crash.

Feynman is probably the most popular physics personality of the modern age. I will quote in later chapters how he mocked the absurdities of postmodern physics, such as string theory.

But, also Feynman did not ponder physics thoroughly enough. He sought to reproduce in his own way much of what the fathers of quantum mechanics had discovered with great effort. However, Feynman was unwilling to deal with the conceptual problems – in keeping with the practical, application-oriented American tradition. Questions about where randomness in nature comes from, or how such a probability function suddenly comes to manifest itself as an electron at a given point, were of no interest to him. Feynman wanted to get rid of the baggage of the musing Europeans, but he made physics too lightweight.

What wonders is the manner in which this theory succeeded for six decades to keep our understanding within the limits set by quantum fi eld theory.[168]
- *Anthony Leggett,* 2003 *Nobel Laureate*

With his disregard for history and an attitude of not taking note of fundamental problems that have troubled Mach, Schrödinger, Einstein, and Dirac, for example, unfortunately it has to be said: Feynman is the founding father of American arrogance[I] in physics.

NO RESPECT FOR HISTORY

Whereas 100 years ago researchers humbly faced the unsolved problems of nature, today students are taught in their first semesters to assume an arrogant know-it-all attitude. At this time, most of them already boast about their guild and how brilliantly quantum electrodynamics is tested.

> Trust those who seek the truth and
> beware of those who have found it.
> - *Attributed to Voltaire*

Feynman instilled in an entire generation a kind of self-confidence that is detrimental to the true researcher of nature. Although he himself drew attention to some weak points in theoretical physics, his attitude created the feeling that everything can be resolved, and that progress is being made. He ridiculed philosophy, which in those years for the first time took on a negative tone, but did not bother to read publications that were a few decades old or even written in a foreign language. He thus touched only superficially on

---

[I] For example, he stated that Einstein had been 'unable to absorb quantum theory' and did 'not understand a thing about nuclei'. But Einstein was only (and for good reason) disinterested in the nuclear models of the postwar period (Davis 1988, p.192f.).

many conundrums with which a generation of deep thinkers had struggled.

About renormalization, which underlies Feynman's quantum electrodynamics, he correctly remarked that it was "probably not mathematically justified," but did not adopt the only honest response, namely to abandon the theory. Feynman apparently believed that the contradictions of his approach would be cured if they were talked about blithely. He thus contributed to the emergence of a culture in which conceptual absurdities were accepted as long as the result of the theory seemed somehow workable. One often hears this being justified by the assertion: "Nature is just like that!" However, this is only modest on the surface: The possibility that one has fundamentally misunderstood something is deliberately ignored.

Feynman described the search for laws of nature as if it was just a matter of playing around. This would have been unthinkable among the leading physicists of a few decades earlier:[169] "In our field we have the right to do anything we want. It's just a guess. [...] experiment may tell us that it's not true." Such phrases would never have crossed Einstein's mind. The credo of Einstein's deductive method was, "I want to know if the Lord God had a choice [in creating the laws of nature]." With his characteristic stubbornness, Einstein did not care how this was viewed by others, but he had preserved for himself a commitment to truthfulness.

Among his peers and followers, Feynman may well still stand out in terms of integrity and sincerity. Nonetheless, he symbolizes the theoretical physics of the postwar era, which, stripped of its roots, was about to take off from the ground of reality.

THE MEPHISTOPHELIAN PACT

After the war, large accelerator laboratories and other large-scale research facilities mushroomed, funded by big money. Their directors were the new masters of physics, although nothing suggested that the successful building of the bomb qualified them in any way for practicing fundamental science. In fact, they did not even know what to do, since the culture of raising fundamental questions was alien to them.

An idea from old Europe (although not a very good one) that could be tested with the big facilities was the neutrino. After the proposal to produce neutrinos with a nuclear explosion had been rejected, it was tried with nuclear reactors. In 1956, over 25 years after Pauli's prediction (perhaps meant only as a provocation), a publication appeared that claimed evidence for the neutrino. A number of methodological shortcomings[170] left many still with reservations about the existence of neutrinos in the 1960s,[171] but these voices have practically disappeared today.

Finally, CERN, the European Organization for Nuclear Research, was founded in 1954. Celebrated as a revival of European research, it was ultimately a copy of U.S. accelerators, even if CERN surpassed them in size and power. Before the war, scientists would not have known what to do with such an accelerator, with the possible exception of Rutherford, who would never have built such a device for its own sake.

> We haven't got the money, so we'll have to think.[I] - *Ernest Rutherford*

[I] Today, almost the opposite is true: as soon as the means are secured, thinking

This was the beginning of a development to build ever-bigger accelerators or colliders, which over the decades grew into a megalomania. Apart from the sheer technical sophistication, which deserves respect (comparable to the architecture of a cathedral), there is hardly any fundamentally new idea in all the particle accelerators that goes beyond the capabilities of the palm-sized cyclotron of 1930, which already accelerated particles on circular paths. And yet, the construction of huge colliders epitomized the American belief that with money, power, and technology, anything could be accomplished.

The light-bulb was not invented by
trying to improve the candle.
- *Michael Faraday*

What has been suppressed is the fact that there are riddles, certainly in the case of the fundamental laws of nature, which cannot be solved using a sledgehammer. *Brute force* does not automatically achieve the goal; sometimes it can even be counterproductive. The large-scale American projects were based on collective efforts that had to be efficiently coordinated. Today's predominant *big science*, that is, research in large collaborations, is the offspring of this research culture that developed in the postwar era.

It is often said that team research is necessary because an individual alone can no longer solve the big problems. This may be true for large-scale experimental projects, but cause and effect are confused here: It is the research tradition of large collaborations that no longer addresses fundamental problems.

---

becomes superfluous.

The only notable pre-war physicist from Europe who participated in this type of research was Werner Heisenberg.[I] Heisenberg, his wife, and some biographers after the war kept emphasizing that he had stayed in Germany only out of a sense of responsibility, despite his opposition to the regime, and that he had made this personal sacrifice in order to help rebuild science in Germany after the war. That may be so, and certainly he was far too cultured to feel any sympathy for the Nazis. Also, he had been viciously attacked by the Nazi ideologues of "German physics."

On the other hand, Heisenberg's ambitions had always been somewhat greater than was beneficial to the pure pursuit of knowledge, even if one leaves aside his involvement in the German atomic bomb project.[172] In the 1950s, Heisenberg once grandly announced a unifying theory with the statement, "Only details are missing," and yet it remained stuck at the wishful thinking stage. Wolfgang Pauli, always in the mood for mockery, then sent him a postcard showing an empty picture frame along with the comment: "This picture shows that I can paint like Titian. Only the details are missing."

A completely sincere attitude towards this new kind of research would have been to clearly state that it contributed nothing to unresolved fundamental questions. Even Heisenberg's proposal to regard the nuclear building blocks of proton and neutron as two states of a single one – later called "isospin" – was ultimately banal and did not add

---

[I] This applied to a limited extent to Pauli, who was also much more critical.

much to our understanding. It actually seems that Heisenberg's best idea dates back to 1925, when he discovered matrix mechanics[I] in Helgoland, which eventually led to his famous uncertainty relation.

One of the Göttingen Eighteen, Heisenberg later opposed the nuclear armament of Germany. As with Carl Friedrich von Weizsäcker, who in 1941 tried to patent a plutonium bomb,[173] genuine love of peace was probably mixed with a somewhat guilty conscience.

With regard to his scientific activities, everything indicates that Heisenberg assimilated quickly to the culture of the new masters and had little interest in the unsolved problems of the 1920s. Not much remained in his work that was reminiscent of the European tradition.

Even Heisenberg's habit of putting forward the beliefs of the majority as the truth was picked up by postwar physics. The competitive spirit in American culture reinforced this trend to marginalize minority opinions, which is bad for science. It is true that Newton and Leibniz were also rivals, as were the Bernoulli brothers or Poincaré and Einstein. But in the end, their common goal was the struggle to discover truth.

Modern research collaborations, on the other hand, often behave like competing companies fighting for market share. Investments have to be profitable, and during the 20th century, Nobel Prizes served as a substitute for knowledge gains, as real knowledge gains could not actually be measured. Often there were trickeries about the priority of a publication and subtle strategies for how to best collect the award.

[I] Although today everyone calculates with the Schrödinger equation.

Science is the search for truth – it is not a game in which one tries to beat his opponent.
- Linus Pauling, Nobel Laureate 1954

A SPECIES BECOMES EXTINCT

What happened to the other leading physicists of the early 20th century? Were they at least consulted as wise oracles on fundamental questions? Not at all. Niels Bohr, whose extensive contemplations on his complementarity principle were indeed attackable, no longer played a role in the postwar period. The same was true for Einstein, who was admired mostly formally. Both his field theory and his criticism of quantum mechanics were met with total disinterest. Schrödinger eventually remained in his exile in Ireland, where he worked on cosmology, on extensions of Einstein's general theory of relativity, and on biology.[1] Working in isolation, he did not receive due respect from the new generation.

Dirac, who pursued his ideas on the *large numbers hypothesis* was held in even less esteem. In later years, he turned to the subject of natural constants and their variability, a very fundamental but grossly underestimated question in today's physics. Dirac died in 1984, having lived for a long time as a scientific recluse.

At the beginning of time, the laws of Nature were probably very different from what they are now. – *Paul Dirac*

---

[1] It was there that he wrote his book *What is Life?*

As different as the fates of Einstein, Dirac, Schrödinger, and Bohr might have been, the isolation experienced by all of them reflects the demise of the European natural-philosophical way of thinking, whose roots reached back centuries, but which no longer bore fruit in the new world of physics.

Chapter 10

Quarks and Neutrinos

Big science at the expense of unsolved problems

The natural philosophical tradition never gained a foothold among physicists in the U.S., and economy of thought therefore did not play a major role in the description of phenomena. To arbitrarily postulate new objects would nevertheless have been unusual at the beginning of the 20th century. But with the newly discovered particles in the 1930s, the floodgates were opened, and the new accelerators in America started an unrestrained "particle production": in 1951 there were about 15 of them, and by 1964 already 75. The array of newfound particles overwhelmed any attempt to produce economical theories for their description. The physicists of the time celebrated each new measurement without giving a second thought to the fact that the number of parameters had long since become reminiscent of the geocentric view of the world. However, the Nobel Committee apparently did not dwell much on the history either and explicitly recognized the discovery of many new particles as "decisive" progress.[174] Needless to say, nobody could explain that these particles were often extremely short-lived, let alone calculate their decay times.

George Gamow remarked[175] that particles were "reproducing themselves like rabbits," and even Enrico Fermi, who was not exactly a man with a profound approach to physics, became thoughtful and remarked as follows in an essay[176] in *Physical Review*, "The probability that all such particles should be really elementary becomes less and less as their number increases," adding scoffingly, "If I could remember the names of all these particles, I would be a botanist." But particle physicists undauntedly penetrated deeper and deeper into the atomic nucleus with ever-higher energies, while the worldview of the European physics tradition dissolved in their hands.

After the thirty fat years at the beginning of the present century, we are now dragging through the lean and infertile years...[177] - *George Gamov, 1966*

MAIN GOAL TO BE ENERGETIC

Robert Hofstadter of Stanford University, for example, fired electrons of record energy at the atomic nucleus, observing that protons did not appear to be point-like. For the huge accelerations that occur in the process, physics has no reliable formula that accounts for the radiation of light; this, too, is a consequence of the unsolved problem of self-energy.[178] In this respect, the deviation from the expected was anything but surprising. Nonetheless, it was deemed fashionable to interpret something new into it and thus the idea arose that protons themselves may be composed of building blocks, so-called "partons". The Nobel Committee continued to honor every result of this kind with awards, and the respective nominees, especially at the top "Ivy League" universities, usually proposed and reviewed each other's work.[179]

When do experimental physicists actually speak of a particle? If highly energetic accelerated particles are made to collide with each other, Einstein's formula $E=mc^2$ states that new particle–antiparticle pairs are created from this kinetic energy itself. For example, at CERN today, a proton accelerated to 1,000 times its rest energy could in theory produce 500 proton–antiproton pairs or even one million electron–positron pairs. The higher the energy, the more conversion processes take place, but this dependence is not a steady one.

At some energies, a particularly large number of particles is produced, which is shown as a peak in the corresponding diagram. These spikes in the diagram are also called resonances, by analogy with mechanical structures, which exhibit strong vibrations at certain frequencies and thereby convert energy. It is not terribly surprising that at high collision energies, these diagrams consistently show zigzags which – regardless of whether their position and size are understood or calculable – are readily interpreted as short-lived new particles prior to decay. One then tries to infer the particle properties from the spatial distribution of the numerous reaction products.

MODERN METAPHYSICS

It is no wonder that numerous new particles were "discovered" through these kinds of experiment and their interpretation. However, the growth of the "particle zoo," as it was already called at that time, was so implausible that people looked for ways to somehow simplify things. American physicist Murray Gell-Mann came up with an ordering scheme consisting of two axes called "isospin" and "strangeness".

It is worthwhile exploring the meaning of these two notions. As already mentioned, the term *isospin* merely sweeps the riddles of beta decay under the rug and sheds no light on why nuclear building blocks of such similar mass as proton and neutron exist, or even why the latter decays after a certain time. The term *strangeness* is even further out of touch with reality. The distinction between "fast" and "slow" decaying particles, associated with the "strong" and "weak" interaction, is superficial in the first place. Now, there are phenomena that break out of this naïve pattern and physicists have therefore labeled them strange, that is, weird. Eventually, then, physicists arrange particles in some kind of diagram, the axes of which might just as well be labeled "incomprehensibility" and "weirdness".

> The most striking feature of the conceptual development of high-energy physics is that it proceeded through a process of modeling and analogy.[180]
> – *Andrew Pickering*

Scholars like Ernst Mach or Hermann Weyl, who only a few decades earlier had dealt with real concepts of nature such as space, time, and matter, would probably have scratched their heads. But Gell-Mann, who designed esoteric patterns in this framework, such as the so-called "eightfold path", is now considered one of the most eminent physicists of the 20th century. For his vague association of this metaphorical arrangement with Hofstadter's hypothetical partons, which were henceforth called *quarks*, he was awarded the Nobel Prize in 1969. Since then, it is generally accepted that a proton consists of two "up" quarks and one "down" quark.

ONE STEP FORWARD, TWO STEPS BACK

It is legitimate to question what fundamental insights about nature were gained by the postulation of quarks. First of all, however, postwar physicists never reflected on the epistemological defeat of decomposing the smallest nuclear building blocks once again. If only the decomposition would work! However, individual quarks cannot be observed, and as an "explanation" for this embarrassment, the theoreticians therefore coined an utterly meaningless notion, namely the "confinement."[I] At that time, nobody seemed bothered any longer by ad hoc postulates whose only value was that they provided an incomprehensible inconsistency with a name.

You cage the mind in a hollow word.
- *Friedrich Schiller*

At that time, it became increasingly clear that the laboratory with the biggest accelerator and detector had the best chance of winning the next Nobel Prize. And in matters of efficiency, the U.S. science organization was far ahead of Europe. At CERN, for example, it was still standard practice until the 1970s for each of several research groups to be allocated only a certain spatial angle from which the collision experiment could be observed. In this case, European thinking in terms of quotas was completely absurd. It was as if 30 competing art historians were each given a square inch of a Picasso painting to comment on.

American particle physicist Burton Richter, meanwhile, was the first to use a detector that completely enclosed the collision point in a way that allowed the created particles to be registered and analyzed from all spatial directions. This earned Richter the Nobel Prize in 1976 for the discovery of

---

[I] In fact, later David Gross designed a model ('asymptotic freedom') that claims that forces between quarks vanish at short distances. Instead, one should have explained why at larger distances the forces became so strong that quarks could not be separated.

the so-called "charm" quark, together with his Chinese colleague Samuel Ting.[I] Naturally, what was measured was just an enhanced energy release ("resonance"), which was interpreted as the formation of a charm–anticharm quark.

...a descending scale of size: atom, nuclei, ... quarks. I cannot suppress the sneaking suspicion that the chain may not end here.[181] – *Emilio Segrè*

A WORLD OF INVISIBLE COLORS

It is claimed over and over again that the quark model has led to an important simplification. If one agrees with the idea that the more than 100 elementary particles discovered up to that point should mean something fundamental, this may seem logical. Meanwhile, the "simplifying" quark model consists of up-, down-, charm-, strange-, bottom- and top-quarks, each with their antiparticles. But the complications did not end there.

One must be careful not to mistake complexity for profundity. – *Karl Popper*

Since the description in terms of quantum mechanics would otherwise have been contradictory, it was postulated that each quark existed in three colors, namely red, green, and blue; however, all three colors had to be included in one nucleon. In addition, so-called "gluons" had to be assumed, which in turn were two-colored. Modern theorizing apparently has no problem in establishing arbitrary rules,

---

[I] There are various whispers about how Richter had outdone the very unpopular Ting in the final stages of the discovery, but one thing is certain: Competition was tough, and all that mattered was to be the first. The noble manner in which David Hilbert, for instance, had refrained from claiming priority for the derivation of Einstein's equations was a thing of the past.

only to furnish them shortly thereafter with similarly arbitrary restrictions. The sole criterion is whether "it works," that is, whether the measurements are described, regardless of whether the description is even halfway rational. With each further complicating adaptation to the measuring results, it is underscored how well the model describes them.

> The agreement of a stupid theory with reality says nothing at all.
> - *Lew Landau*

According to a principle of the philosopher Wilhelm von Occam, the simplest of competing theories should always be preferred. The catchy phrase *Occam's razor* brings to mind a sharp blade to which overly complicated models fall victim. Apparently, in particle physics, it has become blunt over the decades. Sociologist and science historian Andrew Pickering, who has worked as a high-energy physicist for many years, wrote the following in his book *Constructing Quarks:*[182]

> By this stage, the quark-parton model was in danger to becoming more elaborate than the data which it was intended to explain. A critic could easily assert that the sea quark and gluon components were simply ad hoc devices, designed to reconcile the expected properties of quarks with experimental findings.

In any case, there is no question of simplicity, which physicists in the European tradition regarded as a self-evident prerequisite for sound theories.[183] If an intelligent extraterrestrial civilization were to observe us, we must ask ourselves whether it would be smiling at our naïve ideas on nature. Newton, Maxwell, or Planck would quite certainly have done so.

It is the perfection of God's works that they are all done with the greatest simplicity. – *Isaac Newton*

REDUCTIONISM OFF TRACK

Epistemologically, it should be noted that the term "particle" had nothing to do any more with the original idea of the Greek philosopher Democritus (elementary building blocks of nature), since any region in which more energy than usual is absorbed may be interpreted as a particle. The particle is then assigned certain attributes such as *strangeness*, *bottomness* or *hypercharge*, which characterize it like colorful stickers. In the modern way of thinking, a particle is a collection of properties which neither have a systematic structure nor are limited in their number. The term "particle" probably serves more to reassure and supposedly tie in with known things than to denote something real. This is a classic example of nominalistic thinking: Once the concept is established, what it means becomes obsolete.

A word comes along just in time when concepts are lacking.
- Johann Wolfgang von Goethe, Faust.

In this manner, high-energy physics could effectively produce any number of particles. Interestingly, however, the same applies to low-energy physics, which specializes in detecting "particles" of extremely low mass. Since the neutrino experiment of Cowan and Reines in 1956, large underground detectors have been used to record minute amounts of energy that one hopes to identify as particles. Here, too, a major methodological weakness exists, which is related to self-energy or the unknown electromagnetic emissions at strong accelerations. The reason is that during beta decay, in which neutrinos are supposed to be produced, the ejected electron is also accelerated suddenly to a great speed. Although scientists had searched without success for gamma radiation since 1930, it remains at least puzzling why this radiation is not generated in this case.[I]

THE BLOSSOM OF SHADOW PARTICLES

Subsequently, new experiments brought to light further contradictions in a pattern analogous to 1930. Energy was again missing, and usually someone came up with the not exactly original, but no longer frowned upon, idea of suspecting a new particle behind it – hence another neutrino species. From the epistemological point of view, it is particularly risky to take the absence of energy – in other words, the absence of evidence – as positive evidence for a phenomenon. However, such general considerations on methodology are rarely made. Instead, enormous efforts have been made to design corresponding large-scale experiments.

> Neutrino physics is largely an art of learning a great deal by observing nothing.
> *- Haim Harari*

If the experiments of Cowan and Reines on the detection of the particle proposed by Pauli were indeed correct, the mirror-image antiparticle[II] to it had to be produced in large quantities during nuclear fusion inside the sun. However,

---

[I] An X-ray tube, for example, is based precisely on that mechanism of radiation by strongly decelerated electrons striking on metal. An intriguing phenomenon in this context is also the so-called *radiative* decay.

[II] Actually the neutrino itself, while the one detected in 1956 is classified as an antineutrino.

the respective detection reactions were disappointing over many years: there seemed to be way too few solar neutrinos. Finally, the problem was "solved" by introducing another neutrino species, the muon neutrino,[I] into which the missing specimens from the Sun were supposed to have been converted.

THE GODFATHER BATTLESHIP

This theoretical conundrum from low-energy physics led to a symbiotic relationship with high-energy experiments, where the desire to capture the elusive particle soon took hold. In 1963, physicists Lederman, Schwarz, and Steinberger conducted an experiment, using a 5,000-ton steel wall from a decommissioned U.S. Navy vessel for shielding. Such partnerships were not uncommon at the time. On the one hand, they polished up the image of the military as heroes of research and on the other hand, they lent a touch of importance to the scientists who were provided with unique equipment.

Although the article used rather arbitrary, after-the-fact criteria for selecting data and was even ambiguous in its conclusion,[184] the muon neutrino has been considered "established" ever since. In this case, the ambition to discover something theoretically desirable was apparently stronger than experimental care and strictly neutral interpretation, which is what characterizes good science in the first place. Incidentally, this phenomenon has been clearly documented and is known as *confirmation bias*.[185]

[I] This is supposed to be produced by the decay of the muon, itself an unstable kind of heavy electron. The muon had already been registered in 1936 in cosmic rays. Its short decay time of 1.52 microseconds is as unexplained as its mass of 206.77 m_e.

Similar episodes were repeated many times in the following years. Usually, a signal that was too weak was construed to mean that neutrinos had transformed into their cousins, which, of course, had to be "discovered" first. Of course, there is no reason whatsoever why these particles, which are hard to detect anyhow, should behave in such an odd way, except to adapt the theory all over again to contradictory measurements. Why a kind of particle that penetrates the Earth billions and billions of times per second should exist at all remains obscure, too.

> If we are uncritical we shall always find what we want: we shall look for, and find, confirmations, and we shall look away from, and not see, whatever might be dangerous to our pet theories. – *Karl Popper*

SECRECY AND THE EROSION OF SCIENCE

Of interest, but not often discussed, is the connection between neutrino physics and nuclear weapons research. To identify neutrinos, interfering background signals have to be eliminated, to which the uncharged neutrons, produced for example by cosmic rays, also contribute. This is done using theoretical models, which, however, are based on highly naïve assumptions, as nuclear physicist John P. Ralston from the University of Kansas has pointed out. [186] The simulations do not even take into account the characteristics of the detector materials used. One reason is certainly that the reaction behavior of neutrons is a delicate part of nuclear weapons research, which does not exactly help to ensure that the results can be checked by a broad public. Fundamental research and weapons technology are sometimes closer than one might think.[I]

In the meantime, a bizarre state of affairs has evolved in which not only three different types of neutrinos have been postulated (electron, muon, and tau neutrinos), but also a mechanism by which they convert to the respective other form, preferably in such a way that they are not visible. Recently, demands have been voiced to extend this assortment by a fourth, fifth, and sixth kind – so-called "sterile" neutrinos. At the same time, despite enormous technical efforts, it is still not possible to determine the mass of a neutrino. All that is known is that this mass has to be at least one million times smaller than Wolfgang Pauli had originally proposed.[187] This, too, should give pause for reflection. In terms of methodology, one must raise the question of when one should abandon the search for ever-smaller masses, which inherently entails an ever-greater risk of artifacts.

Thus, as technology advanced, two complementary fields of science emerged, and their inherent methodologies ensure that the stream of newly discovered phenomena will never stop, without researchers being aware of this never-ending loop. High-energy physicists can always find something new by further raising the collision energy with new accelerators because incomprehensible phenomena inevitably pop up. On the other hand, the sophisticated techniques of low-energy physics keep lowering the detection threshold, so that the absence of a small amount of energy can always be interpreted as a new particle, too. In this

---

[I] An analogous situation occurs, incidentally, with navigation data from observation satellites, which are important for scientific analysis but are not fully disclosed by NASA for well-known reasons.

fashion, "elementary particles" such as the top quark with 175 gigaelectronvolts and the neutrinos of less than 0.2 electronvolts differ by nearly a factor of a trillion, and we can be sure that it will grow even larger. At the same time, there is still no one who can explain why the only stable particles, protons and electrons, are as heavy as they are.

Beware of false knowledge; it is more dangerous than ignorance.
- *George Bernard Shaw*

FREAK-OUT PHYSICS

By now, people have completely abandoned the idea of trying to predict such mass ratios.[I] Here, too, one has to realize that the manner in which the variety of particles and the quark model were described had nothing to do with the mathematics on the basis of which Einstein, Bohr, or Schrödinger had obtained their results. The domain of mathematics applied to particle description is called group theory. It is certainly interesting to consider, for instance, the group of rotations in a real three-dimensional space,[II] but it has little to do with reality if one considers fantasies such as "isospin" and "strangeness" and rotates them in an abstract space – whatever that is supposed to mean. Wolfgang Pauli therefore referred to this emerging fashion of the 1950s simply as "group plague". Richard Feynman, too, later poked fun at what Gell-Mann grandiloquently called the "unification of interactions":

[I] Because of Einstein's formula $E=mc^2$ physicists use the terms mass and energy almost synonymously.

[II] This is called SO(3). A U stands for two-dimensional complex numbers, so that U(1) is identical with SO(2).

> Three theories… Where does it go together? Only if you add some stuff we don't know. There isn't any theory today that has SU(3) x SU(2) x U(1) – whatever the hell that is – that we know is right, that has any experimental check. Now, these guys are all trying to put this together. They're trying to. But they haven't. Ok?"

Feynman should be reproached for not opposing the arbitrary quark model from the outset. Despite his animosities with Gell-Mann, he apparently avoided open controversy, which, assuming his sincerity, was certainly not a sign of natural philosophical farsightedness. Healthy science also needs tough disputes on the matter, such as those that went on among British physicists of the 19th century about the ether, for example. In addition, the following statement by Ernst Mach has been preserved: "Should these things turn out to be true, I shall not be ashamed to be the last one to believe them."

It seems to me that particle physics is the one area with the greatest unsolved intellectual problems. Some of them are almost a century old, and physicists tend to forget them because there is no inkling of their solution.[188] – *Emilio Segrè*

The notion of symmetry that underlies group theory plays a curious role in modern theories. Since Emmy Noether, David Hilbert's assistant in Göttingen, it has been known that fundamental theorems, such as the conservation of energy or momentum, are related to the symmetry of physical laws with respect to space and time. Consequently, symmetries became a taken-for-granted part of mathematical physics. Yet the extension of these symmetry arguments to "isospin" and "strangeness" is a purely formal analogy, which does not hold any explanatory value yet. In fact, it does not even "work" in the sense of being able to predict particle masses, for example. This leads to the weird situation that the measured masses, which are actually very different, must be "explained" by a "breaking" of the symmetry.[I] The already semantic absurdity of an "asymmetrical symmetry" apparently does not bother anybody in particular.

> ...asymmetry has to be put into the theory and carefully adjusted for no other reason than to produce the desired answer.[189] – *David Lindley*

MATHEMATICAL CHEFS

These thin results were accompanied by a notable arrogance on the part of the then leading theorists at the U.S. universities. Divorced from any historical foundations of their discipline, they thought they could impress nature with their virtuoso calculations. Representative of this hubris is Gell-Mann's own description of his working method:[190]

> ...we construct a mathematical theory of the strongly interacting particles, which may or may not have anything to do with reality [savor this!], find suitable algebraic relations that hold in the model, postulate their validity, and then throw away the model. We compare this process to a method sometimes employed in the French cuisine: a piece of pheasant meat is cooked between two slices of veal, which are then discarded.

It seems that this is rather a matter of displaying one's own wizardry than of a struggle to understand nature. Ap-

---

[I] Symmetry breaking occurs in physics, for example during phase transitions, but this has nothing to do with what particle physicists like to theorize about.

parently, Gell-Mann prided himself so much on his mathematical skills that he ceased to recognize the inherent methodological absurdity in this approach. Although, in contradiction to any European tradition, it is not only a question of whether one considers the idea of explaining nuclear forces in this manner to be nonsense or not

I am mainly annoyed by the flippant attitude of "Yes, maybe our approach is nonsense, but we theoretical physicists are the brightest minds anyway," which surfaces here. One century ago, such a statement would have made any physicist blush with shame. There was a time when researchers considered it a privilege to deal with nature and treated the matter with appropriate earnestness.

It is a great blessing to be one of those people who can and may devote their best efforts to the contemplation and study of objective matters. – *Albert Einstein*

LACK OF MODESTY

The successful physics at the beginning of the 20th century was always guided by the idea of simplicity of the laws of nature, which had been a useful principle since the time of Copernicus. Of course, one can raise the question here of whether nature really has to be simple. Some people object and claim that nature is just the way it has shown itself to be in experiments. The simple picture of a few particles, which dominated the pre-war period, they argue, can no longer be maintained. What does not cross their minds, however, is the idea that the complexity that has emerged could be the result of a lack of understanding. This lack of humility also characterizes the transition from the European to the American tradition. One tries to forcibly impose theories on nature instead of listening to nature attentively.

What is wanted is not the will to believe, but the will to find out, which is the exact opposite. – *Bertrand Russell*

Above all, however, most particle physicists have not yet reflected on the shift in the meaning of the term "particle". There is an underlying misunderstanding that sociologist and science historian Andrew Pickering calls naïve realism. Many mistake for evidence what is essentially an interpretation based on model assumptions.

Pickering observed how a kind of symbiosis evolved between experimenters and theorists in particle physics that favored discoveries. When an unexpected result occurred, this opened up new possibilities: the theoreticians could read it as an indication of a new phenomenon for which there was now experimental evidence. For the experimenters, it was an interesting area that allowed the testing of theories.

Such symbiosis is a far cry from the antagonistic idea of experiment as an independent and absolute arbiter of theory.[191] – *Andrew Pickering*

The fact that longed-for discoveries are more exciting than troubleshooting has always played a role. Checking and considering an error are the frustrating part of the process; expecting a breakthrough, on the other hand, is highly exciting. No human being can avoid this psychological trap, and therefore, in case of doubt, the experiment will always be interpreted in favor of a new discovery, not against it.

But nobody ever won a Nobel prize for proving that something didn't exist or by showing that something else was wrong.[192] – *Gary Taubes*

THEORY OF SCIENCE AND ITS PRACTICABILITY

The Austrian-Jewish philosopher Karl Popper, who, after his emigration in 1936, had to endure the murder of numerous family members in concentration camps, later founded so-called critical rationalism. With the concept of falsifiability, he created a criterion that would define science: A theory is reasonable only if it is able to make predictions and admit failure in the case of their non-fulfillment. Popper's criterion is a sharp knife against ideology and other fantasies, which are all too present in theoretical physics. Yet it does not fully describe the actual practice of the scientific endeavor.

For if a small contradiction appears in an established theory, for example in general relativity, the theory is by no means immediately thrown overboard, especially if no alternative is yet at hand. Rather, auxiliary assumptions are seen to be made in such a case to save the popular theory from being disproved. This has sometimes led to bizarre attachments, often called "standard models." This kind of dynamic was worked out by American philosopher Thomas Kuhn in his 1962 book *The Structure of Scientific Revolutions*. One could say that Popper laid down the laws of how good science should work, and Kuhn described how these laws are bypassed in practice. In this respect, Popper the European and Kuhn the American actually mirror their respective ways of thinking; although Kuhn's observations are, of course, entirely correct. Kuhn is perhaps the most influential science theoretician of the modern age.

Beliefs are more dangerous enemies of the truth than lies. – *Friedrich Nietzsche*

BIG SCIENCE AND GROUPTHINK

The sociological environment in which theories are "established" are usually *big science* collaborations. Strong group identities that fundamentally changed science existed at both Los Alamos and CERN. In these research institutions, and even in entire disciplines, a predominant school of thought often arises. The individual researcher then hardly has the chance to overcome prejudices. By now, the so-called "standard model" of elementary particles has existed far too long for a scientist to doubt its underlying assumptions.

> Few people are capable of expressing with equanimity opinions which differ from the prejudices of their social environment. Most people are even incapable of forming such opinions.[193]
>
> \- *Albert Einstein*

However, overcoming prejudices – as Kepler did when he abandoned circles and realized the elliptical shape of planetary orbits – is usually accomplished by individuals during concentrated work in seclusion.[I] Alvin Weinberg, of all people, the longtime director of Oak Ridge National Laboratory, mused as follows in his book *Reflections on Big Science:*[194] "Discovery is usually an individual act [. . .] I simply cannot imagine the theory of relativity, or Dirac's equation, coming out of teams that nowadays are characteristic of big science."

[I] This is also true in mathematics, where the most important breakthroughs in recent times have been made by mavericks like Andrew Wiles and Grigori Perelman.

Intellectual individualism and scientific
pursuit first appeared together in history
and have remained inseparable.[195]
- *Albert Einstein*

However, such thoughtful voices, which occasionally arose even in the postwar period, had no impact on a seemingly unstoppable science that had perfected the growth and production of results just as Western economies had optimized the manufacturing of consumer goods.

In part because of the huge costs
involved, today government funding is
replacing intellectual curiosity.
- *Dwight D. Eisenhower*

Since the dawn of *big* science, results have been churned out in America's large-scale laboratories as though they were an assembly line. Because of the close intertwining of experiments with theory, this was considered progress, and in a kind of symbiotic relationship, half a dozen Nobel Prizes were generated by respective benevolent peer review. Nonetheless, these activities had little to do with fundamental physics.

Chapter 11

Prestige in Space

Gravity, Rockets, Moon landing

Disgusted by Nazi ideology, Albert Einstein left Germany in 1933 after anti-Semitic forces had been threatening him since 1920. In America, he greatly appreciated the amiability of the people and their uncomplicated mentality. Instead of the roaring, fanatical hordes of the SA, who raged in German cities, he met open and supportive people who accepted everyone as they were.

Yet Einstein's emigration also marked the end of European scientific culture, which even for him could not be restored. Although he was later superficially admired in America, Einstein had no use for the theoretical approaches of post-war physics. With few exceptions, the American physicists were not interested in his thoughts either. In effect, they spoke different languages; Einstein seemed to them a remnant of a bygone era. His isolation was a signal that the European way of thinking would die out with him.

Here at Princeton they think I am an old fool. – *Albert Einstein*

Einstein, like no other, represented the tradition of the natural scientist who relentlessly searches for the ultimate cause. One may assume that he would consequently become a magnet for junior scientists after his emigration, especially at Princeton, where he later received a professorship. No one else had a similar overview of the fields of theoretical physics, some of which had been founded by himself. Instead, Princeton became Einstein's ivory tower.

Certainly, one reason for this was his great obstinacy, which had already made him a "one-horse shay" in Europe, as he called himself. His poor English did not help him to foster contact with the younger generation either. But this does not explain the extent to which his authority was ignored. While he had a few devoted collaborators with whom he still pursued his fundamental ideas, these activities were anything but the center of attention. The extent to which Einstein – after all only 54 years old – was relegated to retirement is striking. In 1920s Berlin, he had been an authority that every leading theorist at least came to consult. However, there is no record that Fermi, Compton, Lawrence, Yukawa, or Rabi ever went to Princeton to discuss an idea with Einstein. They would probably have found little common ground for conversation. America's physics culture, which considered itself young and successful as it increasingly engaged in large-scale projects, was simply not like that. Einstein no longer fitted in.

WHO FORMED EINSTEIN THE ICON?

Much of what has become general knowledge about the life and work of Albert Einstein was influenced by his close friend Abraham Pais – whose biography of Einstein should be taken with a grain of salt, however. Obviously, because of their shared fate as persecutees of the Nazi regime, Einstein felt a strong personal bond with Pais, who had narrowly escaped deportation and death in Holland in 1945. On the other hand, Pais did not really appreciate, indeed could not appreciate, Einstein's approach to physics. A particle physicist himself, Pais later even worked on the outlandish schemes of the Standard Model, the full excesses of which Einstein did not live to see. Moreover, Pais was always disparaging about Einstein's attempts to arrive at a unified field theory.[196]

It is remarkable how biased Pais's approach to general relativity was. He searched Einstein's work for early signs of the formal interpretation that dominated after 1919 and concentrated on mathematical aspects of tensor calculus[I] used by Einstein. This also contributed to the fact that today almost all the work on general relativity is done on the basis of that geometrical interpretation. Einstein's considerations, however, were broader. His very first attempt to develop the theory, which he made in 1911 based on a variable speed of light,[II] is hardly mentioned by Pais, because from his point of view it "made no sense".[197]

The constancy of the speed of light is only valid for space-time regions with constant gravitational potential.
- *Albert Einstein, 1911*

Even when St. Louis-born Robert Dicke developed his own version of general relativity in 1957 with variable speed of light (correcting thereby an error of Einstein), he did not refer to Einstein's work[198] of 1911, of which he was unaware.[III] Despite Dicke's accomplishment, this reflects a

---

[I] The so-*called general covariance* is here frequently mentioned, which refers to transformation properties of matrices (tensors). However, it is doubtful whether this is really an essential part of Einstein's insights. Cf. J. D. Norton, *Reports on Progress in Physics* 56 (1993), p. 791.

[II] I consider this original version to be the most promising; for more details, see *Einstein's Lost Key* (2015).

[III] His collaborator Carl Brans confirmed this to me by email in 2015.

somewhat arrogant attitude that presumes that everything relevant to science has already been published in English.[199]

A HOMAGE YOU DO NOT REALLY WANT

When talking to physicists, one often hears that Einstein spent the last decades of his life dealing with nonsense. This is a statement that reveals the full arrogance of post-war physics; not because every single idea of Einstein was right, but because fundamental questions that Einstein did not cease to ask have been swept under the rug. That aside, Einstein did work with complicated mathematics, but never without demanding to relate to space, time, and matter; by contrast, the supposedly energetic young generation of physicists indulged in abstract, superficial model-building. They resembled teenagers who, together in a group, talk big and believe they can do without the wisdom of their elders.

In America the young are always ready to give to those who are older than themselves the full benefits of their inexperience. – *Oscar Wilde*

There is an astonishing contrast between the blatant disdain for Einstein's way of exploring physics and an ever-increasing glorification of the geometric formulation of general relativity, which has virtually become a religion among cosmologists. This, too, would have offended Albert Einstein, the lifelong searcher and skeptic. Nevertheless, his name has been used with increasing frequency to justify the large-scale projects that were to dominate research in astronomy and cosmology in the decades that followed. Thus, Einstein continues to play a seemingly prominent role in U.S. physics culture of today. However, his work has been stripped of its philosophical content and his ideas have been selectively chopped up as justification for *big science* projects.

BIG SCIENCE IN SPACE

Following Einstein's death in 1955, there was an increasing interest in testing gravitational theories experimentally,[200] which of course is nothing to object to. This led, among other things, to impressive observations such as that of a radar echo reflected from Venus, a test developed by I.I. Shapiro in 1968. When the beam passed near the sun, a slight time delay was observed, in full agreement with Einstein's prediction. In the legendary 1972 Haefele-Keating experiment, the effects of special and general relativity on the elapse of time were measured directly using atomic clocks in airplanes. Einstein was also brilliantly vindicated.

All of this is an undeniable merit of the technology-oriented physics that has flourished to an unprecedented level in the U.S. However, the dream of technical feasibility also led to an eagerness to discover certain phenomena about which Einstein had not expressed himself so unambiguously. This applies to gravitational waves, black holes, and "dark energy." Einstein had first published on gravitational waves in 1917, but his later statements were quite contradictory.[201] For a long time, theorists engaged in a controversial discussion on whether gravitational waves were a prediction of Einstein's theory at all;[202] and if so, whether they would be observable in principle.

Nothing is so difficult as not deceiving yourself. – *Ludwig Wittgenstein*

Large-scale, experiment-dominated physics had always been more attracted by the question of how to measure such waves than by the issue of their existence. This was influenced significantly by the American radio engineer and Navy lieutenant commander Josef Weber, who in the 1960s, after his retirement, embarked on the search for gravitational waves. Weber was a skilled experimenter and brimming with optimism. However, the self-deceptions with which he interpreted random noise as signal also became legendary. A colleague wrote the following about him:[203]

> Joe would come into the lab and turn all the knobs until he had a signal, and then he would record data. . . . Only after that did he define what was to be considered the threshold of noise, and tried twenty different ways of analyzing the data until finally something became visible and he said, "Aha, there you go." Then when somebody came along who knew something about statistics and picked apart his method, he would reply, "What do you mean? When we were looking for radar signals in the war, we tried around until we had it, too." "Yes, Joe, but somebody was sending a signal there." And Joe never understood this.

In this extreme case, science was sufficiently self-correcting to expose Weber's errors. However, gravitational wave research in America grew into a large scientific discipline, whose methods were by no means always coherent. Numerous other reports about the observation of these waves followed, which did not become generally accepted.[204] A general pattern emerges here, namely that the hunt for weak signals is effectively never over if no clean theoretical lower limit is defined of what is expected. When observations fail, models that predict weaker signals gain credibility. Experimenters usually welcome this because it requires building new and even more sophisticated instruments.

Since the measurements that were presented in a major press conference in 2016, gravitational waves are considered established, although there are substantial unanswered questions regarding this issue. Even if one remains skeptical about the results, [205] these laboratories, in the American tradition of postwar physics, have led to the most precise and technologically advanced experiments carried out by humankind.

ADMIRED STAR MONSTERS

Einstein's second thoughts about gravitational waves were one thing, but he was certainly even more skeptical about black holes, the "prediction" of which is often falsely attributed to him.[206] It follows from Newton's law of gravitation of 1687 that masses need to reach a specific escape velocity in order to leave a gravitating celestial body. The smaller[I] and more massive the body is, the higher its escape velocity, which evidently can become greater than the speed of light. Nowadays this is called a black hole, but the concept had already been discussed by the English natural scientist John Michell in 1784. Today's theoretical fashion to treat black holes as mathematical singularities in which all laws of nature would break down[II] was explicitly rejected by Einstein as an improper extrapolation.

However, here, too, a dynamic unfolded which was not uncommon in post-war physics, which would ultimately

---

[I] When moving closer to the center of mass, the gravitational force becomes stronger.

[II] Although the 2020 Nobel Prize was awarded for it, the "proof" that laws of nature break down under certain conditions is somewhat funny - insight is usually gained when you learn that a law of nature is valid.

lead to the recognition of the notion of black holes. In 1967, pulsars were discovered – remnants of stellar explosions that concentrate their mass in a sphere a few dozen kilometers in diameter. They are believed to be giant atomic nuclei consisting only of neutrons and rotating incredibly fast. Among the gravitational physicists, John Wheeler, who had coined the term "black hole", was a particularly attractive figure. Numerous young researchers gathered around him, analyzing the pulsars thoroughly.

Surprisingly, however, there is no specific date to which the first detection of a black hole can be assigned – from a methodological point of view, this is a bad sign. Nevertheless, the interpretation that certain objects might be black holes slowly gained ground, although they could not really be clearly distinguished from neutron stars.[207] Today, there are other observations that indeed suggest very large, concentrated masses in the universe, such as the measurements made on a star orbiting the center of the Milky Way, which were honored with the Nobel Prize in 2020.

Rather dubious was the claim of a group of researchers who publicized their data as a "photograph" of a black hole in the galaxy M87. Although announced in a big press conference, the image consists of numerous theoretical assumptions, while extensive filtering was applied to a confusingly large dataset.[208] It is understandable that people like to interpret such observations as black holes. However, compelling physics is always quantitative, and there is still a lack of evidence demonstrating that the size of these objects really corresponds to the so-called Schwarzschild radius derived from theory.

A PRESUMABLY RELUCTANT CROWN WITNESS

A brazen misuse of Einstein's name occurred in 1998 when cosmologists set out to establish the new concept of so-called "dark energy". Again, technology had provided the stimulus for this. With the Hubble Space Telescope, it was possible for the first time to observe large numbers of supernovae – massive stellar explosions that can be seen over billions of light years. However, the distances measured in this way turned out to be incompatible with the standard cosmological model. The gravitational attraction that had been believed to slow down the expansion[I] of the universe simply did not seem to be present. Without challenging this assumption, a consensus was reached that the missing deceleration was compensated by a corresponding acceleration of the expansion.

> ...the hastiness with which researchers like to believe they have understood a phenomenon, when in reality they have only recorded a description of the facts.[209]
>
> \- *Erwin Schrödinger*

While an explanation for this somewhat far-fetched interpretation was lacking, it was readily postulated that "dark energy" was the culprit that had accelerated the expansion of the universe.[II] As a vague analogy, Einstein's 1917 idea of a *cosmological constant* was called upon,

[I] For these, too, a theoretical explanation is missing until today. A contemplation of this is laid out in my book *Einstein's Lost Key* (2015).

[II] In the meantime, this conclusion has turned out to be premature (https://www.nature.com/articles/srep35596). Still, a sensational discovery receives much more attention than sober counterevidence.

something he himself had called his "biggest blunder". Certainly, it would never have crossed Einstein's mind to solve a contradiction in the observational data simply by postulating an unknown substance. Instead of arbitrarily introducing new parameters, Einstein had always looked for first causes. It is probably not too much to assume that he would have taken the supernova results as an incentive to reconsider the origins of the expansion itself.

MOON BECOMES A MISSION

Key to America's scientific success in astronomy and cosmology, to which many space-based telescopes would later contribute, was the space program. With the 1957 Sputnik shock and the first manned flight in space in 1961, the Soviet Union had suddenly surpassed the U.S. Fears were high of falling even further behind. The Russians had also caught up in the development of nuclear and H-bombs. Combined with the situation in rocket technology, the fear of an attack from outer space was real. At the time, it was considered a military necessity to develop a rocket and space travel program. We can admire the devices that have contributed to the exploration of physics. But it should be kept in mind that physics benefited at the micro level from the development of nuclear weapons, and at the macro level from the technology to get those weapons to their targets. Not exactly fundamental science so far.

The military element was, more or less, conceded by President John F. Kennedy, who in September 1962 announced the Apollo program with the ambitious goal of putting an astronaut on the Moon. He thereby invoked the pioneering spirit of the U.S., which was well received by the public. With the successful Moon landing on July 20, 1969,

America's image, tarnished by the Vietnam War, was also polished up.

> For space science, like nuclear science and all technology, has no conscience of its own. Whether it will become a force for good or ill depends on man, and only if the United States occupies a position of pre-eminence can we help decide whether this new ocean will be a sea of peace or a new terrifying theater of war.[210] – *John F. Kennedy*

FLEXIBLE ENGINEERS

Many Germans had been involved in the development of rocket technology.[211] This was in part a result of *Operation Overcast*, in which scientists and technicians were recruited from Germany after the end of World War II. This resulted in the U.S. securing exclusive military knowledge. Wernher von Braun, for example, had built the V2 rockets for the Nazis, which were still being fired in 1944. The work of Hermann Oberth was also seminal. His dissertation had been rejected as "unrealistic" by the University of Heidelberg in 1922 – one of those instances in which the European mindset was too slow to recognize the potential of the physical sciences. By the same token, the country with the credo of making the impossible possible was just the right place for people like Oberth.

Robert Armstrong's first steps on the Moon, accompanied by the words "That's one small step for man, one giant leap for mankind," which television viewers all over the world followed, also had an enormous cultural impact. The event generated enthusiasm for science and technology and drew attention to the role of our vulnerable planet in the vast universe, a perspective that does not usually dominate the daily news. The physical sciences experienced an upswing, and especially astronomy – an area that had previously not enjoyed particular attention in the U.S. It is said that with his book *Cosmos*, together with the popular TV show of the same name, the astronomer Carl Sagan inspired an entire generation of students to take up astronomy.

Who are we? We find that we live on an insignificant planet of a humdrum star lost in a galaxy tucked away in some forgotten corner of a universe in which there are far more galaxies than people. – *Carl Sagan*

BRUTE FORCE MODELLING

However, this generation was also embedded in a scientific culture where everything seemed solvable by allocating resources and forming large research groups. Solar physics, for example, had a problem with opacity, the impermeability of suns' upper layers to light. In the event that the Sun consists of hot gas or plasma, it is quite inexplicable that it should present itself with such a sharply delineated surface. Finally, in a huge effort of theoretical calculations, this characteristic was described. However, this complicated model required many auxiliary assumptions, which were unrealistic.[212] Therefore, the "Solar Standard Model" is quite similar to the models of particle physics and cosmology, which also originated in the U.S. school of thought. Again, fundamental problems are buried by numerous parameters.

By contrast, Pierre-Marie Robitaille, a former radiology professor at Ohio University, is more "European" in his working style. Robitaille, who in 1998 had developed the world's most powerful nuclear magnetic resonance scanner, shifted his attention to astronomy at the turn of the millennium. He closely studied the genesis of the solar model over the last 150 years, relying on many original publications in Italian, French, and German. Robitaille advanced the revolutionary idea that the Sun consists of liquid metallic hydrogen, a state that can only occur at very high pressure. In 2017, this phase was proven to exist in a laboratory for the first time, although it had been predicted as early as 1935. At that time, however, solar physics had already settled in favor of the gaseous model.

COLLECTIVE SCIENCE – OFTEN IN ERROR, BUT NEVER IN DOUBT

> Even in physics, unfortunately, there are many ideologies [...] Those who do not go with the fashion soon find themselves outside the circle of those who are taken seriously. – *Karl Popper*

Unfortunately, these forks in the road, at which physicists decide by majority vote to follow only one of several paths, are typical of modern science. Since the *community* does not want to live with two conflicting hypotheses, those who hold the minority view are left behind and become marginalized. Moreover, this also has the effect that such rejected alternatives are no longer seriously debated, even in the light of new observations, as is the case with the liquid sun, for example.

> A prejudice is more easily recognized in its naïve, primitive form than as the elaborate dogma it later so easily transforms into. – *Erwin Schrödinger*

Recent images of solar flares, in which material falling back onto the Sun lights up at the moment of impact,[213] cannot in fact be interpreted in any other way than the liquid surface theory. Nevertheless, Robitaille for the moment is still considered an outsider who advocates a supposedly far-fetched hypothesis. The situation is somewhat reminiscent of Alfred Wegener, who had been ridiculed for his theory of continental drift as late as 1928 by a leading geologist in the following terms: [214] "If we were to follow Wegener's hypothesis, we would have to forget everything we have learned in the last 70 years." Wegener's hypothesis, let there be no hiding this, ultimately came to worldwide acceptance through the U.S. military, when Rear Admiral Harry H. Hess published his results on *sea floor spreading*. We can only hope that new solar missions of NASA will also deliver unambiguous results about the solar surface.

All in all, the military technology of space flight also had great benefits for science, which were to become apparent in the following decades in the form of a broad variety of space telescopes.[I] In addition, the reflecting mirrors already placed during the Moon mission made it possible to determine the distance to the Earth's satellite by light travel time. Since then, powerful lasers have achieved an accuracy in the range of centimeters. This not only allowed for precise tests of the law of gravity, but also permitted the calculation of the influence of tidal friction on Earth's rotation. The development of the laser was a groundbreaking innovation in itself, the creation of Theodore Maiman, born in

---

[I] The list of these discoveries is impressive. In addition to the *Great Observatories* that started observing all wavelength ranges from the 1980s, gamma ray bursts were recorded by military satellites as early in 1967, for example.

Los Angeles in 1927. At the time, he was denied funding because, according to the sponsors, his idea had "no practical application."

DAWN OF THE SEMICONDUCTOR WORLD

The direct benefit of manned missions to the Moon was rather limited. That is clear from the fact that after the superpowers' prestigious race to the Moon, there was no reason to repeat these projects for decades. However, the Moon landing has served as a catalyst for many other technologies. The MANIAC[215] computer, built by the Hungarian-American mathematician John von Neumann, had already been instrumental in developing the hydrogen bomb, while the Apollo project would have been unthinkable without the use of computers.

As is known, the first functioning prototype of a computer was built in 1941 by the German computer scientist Konrad Zuse, long before the importance of the technology was recognized in the U.S., where its explosive development started. A major role was played by William Shockley, who developed the transistor, making the efficient application of computer technology possible in the first place. It was Shockley who brought silicon to *Silicon Valley*, which was originally a military research project.

Shockley had been instrumental in both the operational planning of the atomic bomb and the development of U.S. nuclear strategy. Although he was awarded the Nobel Prize in 1956, it later emerged that he had been aware of Julius Edgar Lilienfeld's patents when developing the transistor, but had deliberately concealed them.[216] Lilienfeld, who came from Lemberg, had patented the basic principle in Berlin in 1925. Partly because of growing anti-Semitism, he immigrated to the USA in 1927, where he continued to fight for the assertion of his rights for a long time.

TECHNOLOGICAL EVOLUTION OF CIVILIZATION

Overall, computer technology has left a much more lasting imprint on civilization than space travel. While advanced mathematics came from Europe, computer technology is a typical and almost exclusive product of the American scientific tradition. The basic principles of the computer, like those of the laser by the way, require little insight into fundamental laws of space and time. One might consider the discovery of electrons or the development of quantum mechanics as a starting point, without attributing a decisive factor to them.

Computers were built because hands-on ingenuity continued to look for new applications. With its freedom of research, financial means, and a large pool of talent, America was clearly the best place to do so. Perhaps this also makes it clear that this book is not trying to claim that the European scientific tradition is "better" in all respects. It may be that a greater benefit to humanity (if such a benefit can be defined at all) is found in applications that often require no less creativity and visionary thinking than the study of elementary laws of nature. But historical truths deserve to be told. When thinking about the further evolution of humankind, it is crucial to be aware of the different thinking traditions which, in combination, have led to the level of civilization we have today.

Part IV
From the Moon Landing Downwards – the Degeneration of Physics

Any intelligent fool can make things bigger, more complex, and more violent. It takes a touch of genius—and a lot of courage to move in the opposite direction.

- Albert Einstein

Chapter 12

The Beginning of Gigantomania

High-energy physics devoid of ideas

According to popular belief, science is the exclusive domain of reason. Unfortunately, that is not always the case. The practice of relying on evidence, established since Galileo, has been by and large successful, but it is not infallible. Today, for example, it is becoming increasingly difficult to distinguish evidence from hidden theoretical assumptions. Moreover, what is considered "established" is ultimately a consensus among scientists, in which sociological mechanisms play a major role. Thus, it may happen that even an originally outlandish hypothesis, for which there is insufficient direct evidence, over time is gradually transformed into established truth by targeted research, deployment of resources, the holding of conferences to study it, and so on.[I]

In this fashion, many concepts have emerged in modern physics that have some relation to observations, but still are more an agreement on interpretation than real objects. Such concepts are later no longer even questioned by scientists who pursue alternative approaches. The danger of this happening had already been recognized by Einstein:

> Concepts that have proven useful in ordering things easily achieve such authority over us that we forget their earthly origins and accept them as unalterable givens.

[I] This is often called the `Gold effect' named after cosmologist Thomas Gold, who was the first to correctly interpret pulsars as neutron stars.

The road of scientific progress is frequently blocked for long periods by such errors.

In the following chapters, we shall take a closer look at how physics has justified increasingly absurd theories with ever-more complex experiments. The longer this process continues, the more the watershed between the European and American scientific cultures fades. The latter had practically spread all over the world in the second half of the 20th century, including the countries of the Eastern Bloc.[I] There, no fundamentally different or even "European" research paradigm exists. In these countries, nuclear physics, because of its application in weapons and energy technology, has dominated almost even more than in the West.

In the 1970s, American particle physics was at the peak of its strength and power. The abundant funds available attracted physicists from all over the world to build ever-more powerful accelerators. For decades, the U.S. thus remained the immigration country of choice for the technical intelligentsia, who conducted experiments at increasingly higher energies.

> Physicists are thus led to try safe experiments which are obvious and acquire their interest from the unusual energy region at which they are performed, being otherwise rather simple-minded.[217] – *Emilio Segrè*

Particle physics practiced in this way did not contribute anything to the questions that had been of interest to Ernst Mach or Paul Dirac, for example, but this technological

---

[I] Even though Soviet scientists were often disadvantaged in priority issues if they had not published in Western journals.

branch of science developed a dynamic of its own. Since this was to characterize the high-energy physics of the following decades, it is worth taking a closer look at it. Basically, the idea is always the same – and will always remain the same, even with "new proposals" such as the billion-dollar[218] *Future Circular Collider* (FCC): To smash particles (protons or electrons/positrons) into each other at ever-higher energies.[I] Detectors built up in different layers around the collision point will register a plethora of reaction products produced by the collision energy, in accordance with Einstein's formula $E=mc^2$.

WISHFUL THINKING RATHER THAN MEASUREMENT

By its very nature, it is difficult to catch everything in the process, especially when it comes to uncharged particles that cannot leave any traces in the detectors. Usually, therefore, the absence of energy, if it cannot be accounted for otherwise, is interpreted as a new particle. As soon as the "existence" of such a particle is statistically secured, the experts are persuaded of it, and a Nobel Prize handed out, and the path is clear for the next stage: A new accelerator with even higher energy is constructed.

To interpret the results, all previously found particles are first used. However, higher energies and larger amounts of data typically produce more sophisticated results, so that the existing particles are no longer sufficient to describe them. In this case, a novel particle is postulated, and the cycle repeats itself.

[I] You might also say speed, however, according to Einstein's theory of relativity, at high energies the particle speed is limited by the speed of light.

The novelties of one generation are only the resuscitated fashions of the generation before last.
- *George Bernard Shaw*

A saying that circulates half-jokingly in particle physics is characteristic in this respect: "Yesterday's Nobel Prize is today's background." Yet it reveals a predicament when one observes the evolution over longer periods of time. Many are convinced that the first, often questionable, evidence for a particle had been independently confirmed by subsequent experiments. However, these experiments use the already "established" particle for the description of otherwise unknown signals, the "background." Hence, in reality, the "evidence" is only an interpretation that has been agreed upon. The "established" particles are no longer subjected to a real independent test as soon as the search is on for something new.

To illustrate these excesses, consider, for example, that the "evidence" for the Higgs boson consists in having found more photon pairs than expected in the "background." This background, however, was trillions (!) of times larger than the signal itself. That is irritating, because photon pairs are an otherwise banal phenomenon, which occurs, for example, if an electron-positron pair or any other particle pair is annihilated into energy.

Therefore, what is called the discovery of new particles is not the proverbial needle in the haystack that has to be found, but an additional straw in a heap that weighs millions of tons. It is obvious that here the floodgates are open to systematic errors.

For example, the top quark, whose discovery was agreed upon in 1994 after a lengthy all-night discussion at Fermilab in Illinois, is said to possess an unimaginably short lifetime of only a few 10^{-25} seconds. During this time, even theoretically, it cannot travel farther than one proton diameter (10^{-15} m) from the collision point, let alone ever reach a detector. Thus, the identification again is done only by a cascade of follow-on products and is therefore almost completely based on theoretical assumptions.

On the other hand, high-energy physicists are surprisingly uninterested in the truly visible properties of protons, which they use as projectiles every day. In 2010, for instance, it was found that the proton radius was about four percent smaller than previously assumed.[219] To the best of my knowledge, not a single evaluation routine of high-energy physics has been modified as a consequence of this discovery.[I] This alone indicates how far scientists' working method has departed from reality.

All in all, therefore, these activities must be seen as a hamster wheel from which it is hard for individuals to get off. In fact, many have come to have serious doubts about the meaningfulness of the entire field during their careers. However, such a "cycle," consisting of a project proposal, the allocation of funds, and eventually the construction, takes at least 15 to 20 years. Then, however, these skeptics are usually close to retirement, and the next experiment is again pushed forward by a squad of young, optimistic scientists. That particle physics will ever discover something

---

[I] High-energy physicists would of course object that the radius has nothing to do with the reaction cross section. But this only shows the inability to comprehend the knowledge about nature in logical totality.

fundamental under these circumstances is an illusion, but the belief persists in each respective generation of decision-makers.

The "theoretical" problems that arise during the operation of such an accelerator mostly have to do with the details of the collision process and are utterly irrelevant from a fundamental perspective. It would never have occurred to a European physicist 100 years ago to worry about a "missing transverse momentum," "mono-jets" or a "neutral current."[220] At the same time, already in the 1970s, theoreticians would have been slack-jawed if they had been asked for an explanation of the mass ratio of protons and electrons, the only stable particles in the universe – a problem that bothered all physicists a century ago.[221]

BULLDOZING PHYSICS

The methodological absurdities in which particle physics has gotten lost are, of course, not limited to America, and today there is no research tradition that can be assigned to a particular nationality. The way in which supposedly cutting-edge research is practiced everywhere today merely has its historical roots in the first big particle accelerators of the postwar period. At that time, the structures were created that today still rule the field. The former pure science of physics has entered a state of degeneration.

A look at the way in which dominance was fought for makes this clear. Research institutions used to be run like companies. For example, one of the most prominent directors at CERN was Carlo Rubbia, a vociferous and aggressive science manager from Italy. He undeniably had technical talent, but he cheated, blackmailed, and manipulated whenever it suited his interests.[222]

His numbers are what they are. They are usually wrong. – but if they suit his purpose, nothing is wrong.[223]
- *Bernard Sadoulet*

For Rubbia, it was secondary whether his experiments were unfeasible or his interpretations were wrong; the main thing was that he left the others behind.[I] In 1984, Rubbia received the Nobel Prize, after he had prevented a competing group from publishing in time. The incidents surrounding the so-called discovery of the "W-boson" and the "Z-boson," described by Gary Taubes in his book *Nobel Dreams*, are so disgraceful that a commission of historians should actually investigate this misconduct in detail. Needless to say, Rubbia had neither an idea of, nor an interest in, what building blocks nature is made of and why. The way in which science was conducted at that time can only be called disgusting.

The rationale behind these supposedly scientific data factories is similar to that of the capitalist economy and, like the latter, fails to think in terms of long-term horizons. Investments in big assets must yield their returns in the form of "results," with the Nobel Prize, of course, as the main asset, not because the accolade for scientific insight is valued as such, but because it has increasingly become a prerequisite for government funding.

[I] Leon Ledermann once mocked Rubbia with a parable of two hunters caught by surprise by a bear. "Let's run!" says the one, while the other remarks, "But you can't run faster than a bear!" - "Don't need to. Only faster than you!" was the answer (that characterized Rubbia's behavior).

If one soberly views these institutions as business enterprises that work more or less scrupulously for their own interests, one may find nothing unusual or objectionable about all of this. Civilization is hardly destroyed as a result. However, one must realize that these activities no longer have anything in common with the search for truth to which curious and unselfish individuals have devoted themselves since the time of Copernicus until about 1930. As far as fundamental research is concerned, today it practically no longer exists. This is not because "the simple things have all been discovered," as one sometimes hears. Rather, the failure of modernity is precisely due to the extinction of that culture which went hand in hand with short-term and unsustainable thinking.

Competition, the hallmark of contemporary science, continued. After the European counterpart of the accelerator tradition, CERN, had taken the lead in the early 1980s, the U.S. planned an even larger collider, called the Superconducting Super Collider Laboratory (SSCL), to be built in Texas. When it comes to money, however, politicians also want to know what an experiment will be useful for. Here it is ironic that the anti-intellectual Ronald Reagan, of all people, put the brakes on particle physics.

For the delicate task of explaining the necessity of the machine to Reagan, Leon Lederman was selected, who in the meantime had earned further reputation as the discoverer of the "bottom" quark. "How do you explain particle physics to a president? And above all: How do you explain it to *this* president?" Lederman mused publicly.

In a video made for Reagan, Lederman used a metaphor that was deeply rooted in the American soul: He compared the new accelerator to a cowboy on a path of discovery, alone, westward ... and initially got away with it. In the end, the analogy was not all that wrong; a cowboy is also armed with powerful projectiles, but is certainly without a plan of where to go. From the American point of view, Christopher Columbus was more a scientist than Thales or Aristotle. In this sense, Lederman as a scientist was not unlike Reagan as statesman. Eventually, however, the U.S. Congress cut off the funds.

For the next step, still greater efforts will be required, without the guarantee that it will be the last step. The limit would be set by the cost, without the guarantee of ever reaching the end.
- *Emilio Segrè*

A FIG LEAF FOR THE MASS PROBLEM

Due to the cancellation of the American collider project, it took almost 30 years until CERN had again built a functioning device, in the form of the Large Hadron Collider. As the desired result, this time the so-called "Higgs boson" was chosen, a 50-year-old idea of a humble Scottish physicist. Of course, the fashion that was established among theorists after the war to describe all processes in terms of particles, which Peter Higgs also followed here, is altogether nonsensical. Above all, however, one should not indulge in the illusion that these theoretical concepts somehow have a concrete link to experimentation. As previously, it was enough to find a cluster of events in an arbitrary energy range (there were no predictions[I]). Then this was associated with the idea of Higgs, with the metaphorical considerations about group theory mentioned above.

It is often parroted without reflection that the Higgs boson explains the mass of elementary particles. This lacks any foundation. There is not a single particle mass which the Standard Model could calculate, not even a mass ratio. What we have here is no more than a vague association to an abstract mechanism.

The last physicist who had reflected seriously and fundamentally on particle masses was Paul Dirac. By the way, it follows from very simple reasoning by means of physical units that the phenomenon of mass in physics cannot be explained without a gravitational constant.[224] This, respectively the origin of gravity itself, can be understood, if at all, with distant masses in the universe, as the mentioned Viennese physicist Ernst Mach had suggested in 1883.

The fact that modern publications on the issue of mass often do not even mention Ernst Mach must baffle every educated physicist in history. Particle physicists even dispute that gravity plays an important role in the forces of nature; since it is so weak, they argue, it can actually be neglected. In all seriousness, one can read this in a textbook about quantum field theory.[225] These "communities" have completely lost the ability to think outside their box.

[I] Sometimes it was called a 'prediction' that people had already searched in all other energy ranges without success.

THE WISHFUL THINKING JOURNEY INTO THE UNIVERSE

Some particle physicists, at least intuitively, have realized that the gigantomania of accelerators would have to come to an end at some point, and in time they managed to find a new foothold. Steven Weinberg, one of the most influential physicists of the late 20th century, received the Nobel Award in 1979 for the so-called weak neutral currents. From a fundamental perspective, the phenomenon is as special as it is irrelevant, but over the years it has been hyped and, with great effort, eventually read into the observational data. It is often called "evidence" for a unification of the electromagnetic and the "weak" interaction. The latter is, as mentioned, merely an incomprehension of beta decay dressed up in formula language.

In any case, it had become clear to Weinberg that the future also lay in astrophysics and cosmology (the first space telescopes had already been launched), and so he published his book *The First Three Minutes*, which was meant to describe the primordial era after the Big Bang. Weinberg can be credited with having stimulated a general interest in cosmology. However, his observations were by no means accurate enough to justify the title of his book in any way. Such exaggerations do a disservice to science.

Speaking generally, such extrapolations, that is, assumptions that go far beyond the tested range of the validity of a theory, indicate not only a lack of scientific judgment, but also the hubristic way in which people have approached natural science research. Such an attitude would have been incompatible with the scientific ethos that still prevailed around 1930. Since Hubble's observations, an expansion of the universe was assumed, but to extrapolate back to the Big Bang is an excessive step – as if one wanted to predict the course of an entire movie from a snapshot. If at all, the earliest reliable data are witness only to a period about 400,000 years after the Big Bang;[226] for earlier epochs, the view is irretrievably obstructed. Trying to deduce from these measurements what is supposed to have happened in the first minutes, seconds or whatever fractions of a second after the birth of the universe, is like reading tea leaves.

Nevertheless, today's physicists indulge in the wildest speculations about which elementary particles might have existed at that time, a bad habit to which Weinberg has contributed a great deal. Professional astronomers, more aware of the measuring precision of their own observations, were often annoyed by this. For example, Michael Disney[I] of Cardiff University scoffed at particle physicists who, having had their field "paralyzed by its escalating cost, have moved over into cosmology, wishfully thinking of the Universe as 'The great Accelerator in the Sky'."[227]

Nonetheless, Weinberg remained an influential trend-setter for the unification fantasies of particle physicists and cosmologists that became popular in the 1980s. In his book *Dreams of a Final Theory*, Weinberg is amazingly uncritical about the expectations of some theorists, who at that time promised everything under the sun about grand unifications without being nearly as self-critical as, for example, Einstein and Cartan in 1930. Nothing of what was written in the 1980s about the physics of the very largest and the

---

[I] The discoverer of *Low surface Brightness Galaxies*, a particular enigmatic type of galaxies.

very smallest was anywhere close to the depth of those earlier considerations.

HUBRIS TO THE POWER OF TEN

A heroic age in which physics temporarily broke away from its experimental anchors...[228]
- *Steven Weinberg*

While Weinberg was more cautious when talking big and did not want to mess with any of the emerging theoretical fashions, his colleague Alan Guth clearly crossed the line to megalomania in his work. The following quotation speaks for itself:

> The mystery of creation is not such an unsolvable riddle any more. We now know what happened 10^{-35} seconds aft er the Big Bang. [...] These spectacularly bold theories attempt to extend our understanding of particle physics to energies of about 10^{14} gigaelectronvolts – absolutely fantastic.[I]

At this time, physics finally bade goodbye to the humility that had characterized Einstein, Bohr, and Schrödinger. Guth actually speculated about extrapolating the already unreliable theories about the first seconds after the Big Bang back by another forty powers of ten. The idea that such an approach will ever be testable by observation is completely ludicrous. Tellingly, however, Guth received about a dozen job offers from American universities for his vague idea, later called "inflation," before accepting a professorship at the prestigious MIT. This alone demonstrates that, already in 1980, theoretical physics was sick.

[I] The heaviest stable particle, the proton, corresponds to about one gigaelectron volts (GeV), thus 10^{14} GeV would be about ten billion times more energy than the LHC at CERN currently produces.

Nothing hinders the progress of
science more than thinking you know
what you do not know yet.
- *Georg Christoph Lichtenberg*

In the American thinking tradition, there is a sympathy for bold visions. Indeed, in applied science, this has led to the greatest technological achievements of mankind, in which European thinkers have often failed. The difficulties in such cases usually lay in practical implementation, while the theoretical principles were clear. Yet fantasizing about theories as such, the practical test of which is impossible from the outset, can hardly be called bold. It is simply out of touch with reality.

CONTEMPORARY BLINDERS

In addition to the above, Guth's proposal exposed his historical ignorance, since he was obviously unaware of the respective studies of Erwin Schrödinger, Albert Einstein, and Ernst Mach. Within the picture of an expanding universe, it was disputed for a long time whether the expansion would continue forever or be slowed down by gravity in such a way that it eventually contracted again. Oddly enough, observational data indicated that the cosmos was mysteriously fine-tuned between these two possibilities – a profound enigma. In another context, however, this had already been noted by Schrödinger[229] in 1925, which suggests that the cause of gravity lay in the distant masses in the universe. It was this idea of Ernst Mach that the American physicist Robert Dicke had taken up in an article[I] in

[I] Cf. chapter 11. Likewise, see a 1953 essay of the British-Egyptian cosmologist Denis Sciama (MNRAS 113, p. 34).

1957. Dicke ingeniously combined this with a theory of variable speed of light, without knowing about Einstein's idea of 1911, however. Guth, on the other hand, merely mentioned in his publication a “riddle of Dicke” (which in reality was one of Mach) and went on to propose his “solution” that the universe was “inflated” with superluminal velocity during its very first fractions of a second. What nonsense.

It is amazing that an idea as banal as it was untestable could survive in science. In his book *The Road to Reality,* the British mathematician Roger Penrose carefully prefaces his criticism with the following:[230] “Since I believe that there are powerful reasons for doubting the very basis of inflationary cosmology, I should not refrain from presenting these reasons to the reader”, before thoroughly ripping the concept apart. However, the following assessment of Penrose is also documented: “Inflation is a fashion high-energy physicists visited on cosmology. Even aardvarks think their off spring are beautiful.”

In the U.S. tradition, one generally speaks of a model as if it were an aircraft design. Just as one would tweak wings and rudder until the flight characteristics satisfy, one can – according to the paradigm – fit certain details of nature in a model within a theoretical framework. There is no limit to the number of adjustable screws. Obviously, this notion of a descriptive *model* is diametrically opposed to that of a *theory* in the tradition of pre-war physics. At that time, the aspiration was to explain all properties of nature without making any further assumptions or using arbitrary parameters.

CALCULATE, WHATEVER IT COSTS

> Whether... the indispensable contact with the sensory world is sufficiently preserved. Without such contact, even the most perfect world view would be nothing more than a soap bubble, which bursts at the first blow of the wind.[231] – *Max Planck*

A characteristic feature of the working style of theoreticians that emerged in the 1980s was complex calculations that were highly challenging from a technical point of view. Increasingly, the benchmark became whether the calculations were internally consistent, regardless of whether there was any contact with reality. The kind of problems that were being described sprang from the metaphorical concepts of particle physics that allowed all kinds of mathematical associations. However, it is important to keep in mind that this custom of theoretical calculations had nothing to do with what was once considered mathematical physics in Europe.

The function spaces of quantum mechanics, nonlinear continuum mechanics, the differential geometric formulation of general relativity, or its extensions[232] that Einstein discussed with Élie Cartan around 1930 are highly sophisticated branches of mathematics. However, practically nothing of what has been calculated since the 1970s is based on them. Logically, therefore, none of the serious problems that existed at that time has been solved. Lee Smolin, one of the few contemporary physicists to criticize today's methods, wrote the following:[233]

> This style is pragmatic and hard-nosed and favors virtuosity in calculating over reflection on hard conceptual problems.... however, the lesson of the last 30 years is that

> the problems we're up against today cannot be solved with this pragmatic way of doing science ... to put it bluntly, we have failed.

Since then, another 15 years have passed without any fundamental progress. This failure has multiple causes, which can only be understood by knowing history. It is not only about physics being in a phase of "normal science," in which experimental results are pigeonholed without thorough reflection on the method; in addition, a system of institutions has emerged to ensure that these self-imposed limits of reflection are not breached by established science.

Chapter 13
Fantasies of Omnipotence
Strings, Multiverses, Supersymmetry

Over there is not the place to learn
modesty and restraint.
- *Theodor Fontane*

No matter how one may assess the endeavors of theoretical physics after World War II, one thing is certain: they were completely detached from the way the laws of nature had been examined in the past. The problems were completely different, but so were the methods employed. In this respect, it must be considered completely naïve to believe in progress without first having solved the problems that were already known in 1930. Without taking care of those problems, theoretical physics then turned even more abstract in the 1970s and its connection to reality became increasingly less recognizable.

The erstwhile European tradition of tackling these problems with profound reflection, exemplified by the mindset of Ernst Mach and Albert Einstein, for example, might be called *Think Deep*. By contrast, *Think Big* is a slogan of U.S. culture. Correspondingly, in today's physics, the idea of a large-scale, bold theory design always holds a positive connotation: the more daring, the better.

If a false thought is so much as
expressed boldly and clearly, a great
deal has already been gained.
- *Ludwig Wittgenstein*

Tellingly, many theories of that time used the attribute *super* for themselves, for example "supersymmetry," also called SUSY. However, apart from the name, this symmetry has little to do with the symmetry theorem of Emmy Noether, who based it on real concepts such as space and time and not on exotic properties that were later created by particle physics.

If there is something in "supersymmetry" that has to do with *super*, it could be "superficial." Supersymmetry is based on such a superficial concept of particle physics. Undeniably, there are two different ways in which matter behaves on a microscopic scale. So-called bosons with integer spin – a hydrogen atom, for example – can theoretically permeate each other at will and thus concentrate in a very small volume; something that has meanwhile even been proven experimentally.[I] Fermions with half-integer spin, on the other hand – electrons, for example – are not allowed to do so. A "gas" of electrons is therefore much harder to compress. A compelling reason why nature adopted this distinction is not known to this day; nor is it known why nature exhibits such a strange phenomenon as spin in the first place. Without solving this underlying riddle, notions built upon it, such as fermions and bosons, ultimately remain naïve. All the evidence suggests that on the microscopic scale, the concept of particles does not properly describe reality. The "modern" idea of particles as a collection of arbitrary properties has in any case moved even further away from reality.

[I] In 2001, the Nobel Prize was awarded for the experimental proof of Bose-Einstein condensation, first predicted in 1924 by the Indian physicist Satyendranath Bose.

To "explain" the existence of fermions and bosons, supersymmetry postulated that for each type of particle, a corresponding partner of the other type must exist. There is not the slightest evidence for this, but one would expect that the twins constructed in this way would at least share the same properties – an identical mass, for example. Since this is obviously not the case, the "SUSY" theory assumes the symmetry is "broken" again in this case (and thus asymmetrical after all), which leads to a much higher mass of the not yet discovered partner. As earlier, this "broken" and therefore unfortunately asymmetric symmetry has been justified with a vague analogy to the physics of complex systems,[I] which has nothing whatsoever to do with it.

Nevertheless, physicists have expended great effort to find supersymmetric particles. Since the first experiments in the mid-1970s, the hunt for these particles has been a protracted story of failure. The lingering methodological problem is that supersymmetry does not provide an upper limit on the masses of the particles it invents. With this unhealthy flexibility, any result[234] could be "explained" – if indeed there happens to be one. Over the past 40 years, this has led to a situation where different variants of the theory made predictions, which were then tested in a particle accelerator. Once nothing showed up, the theoretical calculations were adjusted, which in turn motivated the construction of a new machine. This continues to the present day.

[I] For example, a phase transition like steam to water. This is called 'spontaneous' symmetry breaking, an effect of complex systems that has no visible connection to particle physics.

A great effort, shamefully! is wasted.
- *Johann Wolfgang von Goethe*

The most recent failure took place at the Large Hadron Collider (LHC) in Geneva, where again the hoped-for SUSY particles refused to be discovered. Some, like Nima-Arkami-Hamed, one of the most frequently cited contemporary physicists, were nevertheless brazen enough to claim after the fact that the "best people" had not expected a confirmation of supersymmetry. Sabine Hossenfelder, author of the book *Lost in Math*, pointedly commented as follows:[235]

> But not one of those "best people" spoke up and called bullshit on the widely circulated story that the LHC had a good chance of seeing supersymmetry or dark matter particles. I'm not sure which I find worse, scientists who believe in arguments from beauty or scientists who deliberately mislead the public about prospects of costly experiments.

Yet the business keeps going unswervingly. In 2021, the 28th conference "Supersymmetry and the Unification of Fundamental Interactions," hosting more than 1,300 participants, took place – as always, with grand promises and zero results.

History suggests, however, that if these superparticles don't turn up, there will be strenuous eff orts to save supersymmetry by tinkering with it rather than deciding that the whole thing is a failure.
– *David Lindley, in 1993*

UNDEAD RESEARCH

Science critic Bruce G. Charlton coined the term "zombie science" for such kind of activity. Under normal circumstances, a nonsensical hypothesis would be viewed as disproven if it failed to predict reality. However, if one allows for endless evasions in secondary hypotheses, they "explain" why the failure to describe the facts has not yet led to the death of the theory.[I] Charlton calls such research traditions – which are kept alive only by continuous transfusions of money – the "walking dead" of science. Supersymmetry is a textbook example of what science theorist Imre Lakatos called a "degenerative" research program: one that is undertaken whenever new auxiliary assumptions are needed to cope with the problems of incomprehensible observations.

This decline of theoretical physics has been called out by Roger Penrose in his book *Fashion, Faith and Fantasies*, yet the part of physics that deals with experiments and observations is not necessarily healthier. After all, the decades-long failure of the theorists when making predictions has always been a welcome occasion for experimenters to call for the construction of new facilities.

What is to the mutual benefit of theorists and experimenters, however, is to the detriment of society in two respects: as a waste of resources and, perhaps even worse, as an intellectual stalemate hiding behind fantasy research. Theoretical physics has become science fiction with equations; the collider experiments science fiction with big machinery.

[I] A great survey of the history of these scientific prevarications is given by David Lindley (1993), while Sabine Hossenfelder (2018) describes in detail the most recent attempts involving the 'minimal-supersymmetric' Standard Model.

RETIREMENT PROTECTS FROM REFUTATION

The long time frames involved had the effect that the respective researchers lost credibility due to their false predictions. In the meantime, however, a new, confident generation had grown up, which was pretty confident that *this* time it would work. Richard Feynman once scoffed at such optimism when talking about a similarly ambitious idea:[236]

> Somebody makes up a theory: The proton is unstable. They make a calculation and find that there would be no protons in the universe anymore! So they fiddle around with their numbers, putting a higher mass into the new particle, and after much effort they predict that the proton will decay at a rate slightly less than the last measured rate the proton has shown not to decay at.

Needless to say, this could only be checked by new experiments... Here again, one encounters the typical symbiosis between theorists and experimenters in post-war *big science*: In the end, both benefit from the postulation of something unknown, whose limits of testability were never precisely defined. Early 20th century physics was different: At that time, it would not have crossed the mind of any experimenter to build an expensive apparatus based on a guess of the theoreticians. They had enough ideas of their own not to merely scale up the previous setup. By the same token, theorists were busy with things that had been found experimentally, not with stuff they only hoped to find.

Frequently, the contemplation of supersymmetry was also motivated by astrophysics: There seems to be insufficient mass in galaxies, especially at their edges. This inferred from the not diminishing velocities in the outer regions (technically called *flat rotation curves*), which is why "dark matter" has been postulated. Ultimately, however, the main thing that particle physicists and astrophysicists have in common is that they would love to see things that are not there. Alas, the non-observation of something is simply evidence for nothing at all. People who are knowledgeable about galaxy dynamics doubt, by the way, whether supersymmetric particles can explain the phenomenon of "dark matter", which poses many conundrums that are unknown to the majority of particle physicists.[I]

> Physicists know that the rotation curves are flat. They don't know anything about the regularity of rotation curves or global scale relations and aren't very interested in learning about them. – *Robert Sanders*

Since the dark matter phenomenon occurs only at very low accelerations in the order of c/T_u (T_u being the age of the universe), it is not unlikely that there is a fundamental reason behind it, something most physicists prefer to gloss over.[II] The Dutch radio astronomer Robert Sanders comments as follows in his book on dark matter:[237]

> The real problem is that dark matter is not falsifiable. The ingenuity and imagination of theoretical physicists can always accommodate any astronomical non-detection by inventing new possible dark matter candidates.

What is falsified, on the other hand, is the standard model of cosmology itself; according to Pavel Kroupa of the University of Bonn the distribution of the dwarf galaxies

---

[I] These include various strange coincidences, which rather indicate that the law of gravity must be modified. See in this context also M. Disney: "Galaxies appear simpler than expected", www.nature.com, 23.10.2008, See also Sanders (2010), p. 167.

[II] There are a couple of interesting approaches, such as *quantum inertia*, that relate these accelerations to cosmological data, as proposed by British physicist Mike McCulloch (2014). Another theory became known as MOND (Modified Newtonian Dynamics). The mainstream, however, prefers to continue undisturbed by such deeper reflections.

around our Milky Way is incompatible with this model.[238] Nevertheless, today's theorists of cosmology are busily juggling with the invisible substances of "dark matter" and "dark energy," especially in numerical simulations that fit the models to the observations.

People take the unexplained dark thing
more seriously than the explained
bright one. – *Friedrich Nietzsche*

NEW IDEAS, THE ELIXIR OF IMMORTALITY

An attempt similar to SUSY in terms of language and mindset is the theory of "supergravity." Although nothing supports the fact that gravity can be described with the help of particles, the idea of "gravitons" or even "gravitinos" has appeared in connection with the supersymmetric theories. In his inauguration lecture in Cambridge in 1979, Stephen Hawking, who at that time was already used as a public oracle of physics,[I] expressed the hope that with supergravity, the "end of theoretical physics" might be reached. In fact, there was a sense of euphoria among theorists because they thought they could solve the problem within months or even weeks. This psychological state of the *community* actually speaks for itself; it can only be called a loss of reality. The hallmark of this illusion, by the way, was a working method of abstract, excessive calculations, diametrically opposed to that of Einstein, who usually gained his insights in an intuitive way from basic principles.

[I] It is barely imaginable that the severely handicapped Hawking has written his numerous books himself. In this respect, what is published under his name is to be taken with caution, not only because of the excessive speculations that occur in the later work.

True witchcraft calculation, sufficiently protected by its complexity from being proven false. – *Albert Einstein*

By this time, such more general considerations were already explicitly discouraged. *Shut Up and Calculate* became the dictum of the theoretical method, and perhaps it was for this reason that almost unlimited resources in the form of mathematicians were available. The problems of physics could simply be hammered out by using the necessary resources – so the thinking went. However, these dreams were soon dashed, and the theory of supergravity was buried without much ceremony. Science author David Lindley sarcastically commented on this as follows:[239]

> It is hard to find an obituary for supergravity in any scientific journal. No doubt some physicists paused briefly to reflect on the passing of what had seemed a wonderful idea, but no one took the time to eulogize. Scientists are not inclined to pause and reflect at the demise of a once-bright idea; if they experience any grief, they find release from it in action, not in mourning. Their idea of recovery is to move as quickly as possible to the next bright idea. And, as it happens, a new idea - another superidea - was waiting for them.

THE MOTHER OF ALL HOPES

The next object of enthusiasm became string or superstring theory, which assumes that elementary particles can be described as tiny vibrating strings. Since point particles had led to contradictions time and again, this hypothesis of small strings is not as original as it is often portrayed. Moreover, these strings are hopelessly unobservable. Their assumed size of 10^{-35} meters would be smaller than the atomic nucleus (10^{-15} m) by the same factor the nucleus is smaller than the Earth. Given that technology already faces principal barriers at this scale, the idea of ever finding clean evidence for strings is laughable.

However, the unwavering optimism with which these theories were worked on in the American tradition is quite remarkable. Although Einstein was also cautiously confident about his unified field theory, such a full-throated, beyond-all-doubt attitude of soon being able to solve the final mysteries of nature would have been unthinkable in his time. Here are some examples.

Murray Gell-Mann praises the virtues of string theory in his book *The Quark and the Jaguar*, musing about Einstein's declining mental capabilities when the then 50-year-old was dealing with unified field theory. Apparently, Gell-Mann missed the irony that he had published his own book at the age of 65. In 2001, David Gross, a subsequent Nobel laureate, gave a lecture entitled *The Power and Glory of String Theory*, although the theory had not explained a single measuring value so far. Nothing indicated that string theory had anything to do with reality, yet it became the Promised Land of theoretical physics.

"No one could have imagined in his wildest dreams that we would get where we are now" stated Brian Green of Columbia University in New York, for example, not being afraid to draw big comparisons:

> According to superstring theory, the marriage between the laws of large and small is not only happy but inevitable... String theory has the potential to show that all of the wondrous happenings in the universe... are reflections of one grand physical principle, one master equation.

Also others have been captured by the idea of a "mother of all equations".[240] Coincidentally or not, "Mother of all battles" and "mother of all bombs" were, also common jargon at that time. The fact that string theory has failed to produce even a single equation containing physical quantities seems instead to be the mother of all embarrassments.

But ye only prate, and are the bell of your doings. – *Friedrich Schiller*

EUPHORIA VS SKEPTICISM

Michio Kaku, a professor at the City College of New York, speaks even more exuberantly: "String theory is a 'higher' theory than Einstein's, it will unravel the secret of the Big Bang, what happened before Creation, what happened before Genesis 1:1: namely, that our universe came into being after a collision with other universes—all of which string theory allegedly predicts."[241] One thing you can observe with these guys is that the more they talk big, the more religious their wording sounds. They can hardly be distinguished from the other sectarian preachers who can be found in abundance in the U.S.

We are more naïve than those of the Middle Ages, [...] for we can be made to believe almost anything. – Neil Postman

Some physicists are embarrassed by Kaku, who likes to be seen with an Einstein hairdo, yet open dissent to his ramblings remains rare. The empty promises of string theory nonetheless led to criticism even from physicists who were basically optimistic, including an "architect" of the Standard Model, Nobel laureate Sheldon Glashow. He complained that string theory was not physics and compared its bold assumptions to scholastic debates about how many angels could dance on a pinhead. His colleague

Gerardus 't Hooft mocked the promises of string theory as follows:[242]

> Imagine that I give you a chair, while explaining that the legs are still missing, and that the seat, back and armrest will perhaps be delivered soon; whatever I did give you, can I still call it a chair?

Richard Feynman was also quite blunt about what he thought was a mirage:

> I don't like that they're not calculating anything. I don't like that they don't check their ideas. I don't like that for anything that disagrees with an experiment, they cook up an explanation – a fix-up to say: "Well, it still might be true". For example, the theory requires ten dimensions. Well, maybe there's a way of wrapping up six of the dimensions . . . but why not seven? When they write their equation, the equation should decide how many of these things get wrapped up, not the desire to agree with experiment. [...] String theorists don't make predictions, they make excuses.

HERD INSTINCT AND PROPAGANDA

It is certain that among the numerous string theorists – more than 6,000 people are said to be working in the field – there are also many excellent mathematicians. However, nothing indicates that they have taken the effort to understand physics in its historical development, but rather build on the models of American particle physicists from the 1960s, without any further reflection. Sometimes it is argued that the Standard Model should be extended to a "complete theory" by strings; on the other hand, it is also argued that string theory already contains the results of the Standard Model. From the historical-methodical perspective this is ridiculous. Newton's theory of gravitation, which constituted real progress, did not simply not "contain" the medieval epicycles; it consigned them to the junk room of history. As long as the Standard Model does not occupy its due place in that junk room, there will be no real progress.

If science is seen as an instrument for shaping the future, such a combination of high intelligence and historical ignorance as in the case of string theorists is by no means innocuous. It is precisely because the string theorists are so completely divorced from the traditions of Einstein, Schrödinger, and Dirac that they have lost touch with reality. With the removal of philosophy from physics, a supervisory instrument has been lost – one that would have opposed these brilliant but mindless calculations. Striving to be part of an intellectual elite is not at all helpful if one really wants to discover basic laws of nature. Rather, it has led to a downright grotesque arrogance as far as the status of string theory in physics is concerned. String theory, as many say, is *The Only Game in Town* – the only reasonable way to look for a fundamental theory.

The customs of your tribe are not laws of nature. – *George Bernard Shaw*

One is inevitably reminded here of American exceptionalism, which of course presumes that Western political systems are the only true representatives of democracy and human rights and must be exported – if necessary, by means of *regime change* and bombs. With a similar attitude, today's theoretical physicists also doubt the legitimacy of an idea if it is not based on string theory.

Blind commitment to a theory is not an intellectual virtue: it is an intellectual crime. – *Imre Lakatos*

Notwithstanding all the legitimate criticism of the Western democracies, they at least have certain objective merits.

String theory, on the other hand, has not yet proven its worth in any field and can therefore only be called a scientific ideology. As in politics, a faithful group of followers can nevertheless be created by appropriate propaganda.

Nothing is more dangerous than an
idea, when it is the only idea we have.
– *Emile Alain*

IN THE LAND OF UNLIMITED ILLUSIONS

I do not want to tire the reader with the terminology of string theory, but one could perhaps use the internet to verify for oneself how much research activity is devoted to certain concepts that are obviously devoid of any sense. Much ado has been made about a so-called "duality" in string theory. This is some sort of quantum field theory (which is a questionable concept in itself, cf. Chapter 9) thought to have some similarity with a type of quantum gravity (admittedly, no such thing exists yet). The same is true for the so-called "AdS/CFT correspondence," about which there are thousands of publications.

These vague parallels of half-baked concepts were unashamedly called the "second revolution" of string theory, while the term "first revolution" had already been reserved for the theory as such. Strictly speaking, it is not even a theory – it is much too sketchy for that – but a collection of an absurd number of theory versions. These 10^{1500} (!) possibilities are known as "landscape" and associated with "parallel universes." Some claim that in these universes, other values for the enigmatic fine-structure constant $\alpha \approx 1/137$ exist – for example, 1/138 in the neighboring universe etc. – thereby solving the problem. I admit I do not know the best explanation for this constant, but parallel universes are certainly the dumbest one.

The early promise of superstring theory to calculate these quantities has faded after decades of disappointing experience in attempts to construct phenomenologically adequate solutions, together with the discovery of multitudes of theoretically unobjectionable but empirically incorrect solutions.[243] – *Frank Wilczek*

So why haven't all these extravagances already led to public debate? The reason is that theoretical physicists are regarded as extremely smart and it is considered natural not to understand the things they are dealing with. Moreover, string theorists like to subtly state that if you are unfamiliar with their subject, you are not entitled to criticize them in a proper way. The clerical intelligentsia in the Middle Ages might have forbidden criticism with the same argument. However, you do not need to have studied theology to be permitted to demand evidence as a proof of God.

If string theorists are wrong, they can't be just a little wrong… then we will count string theorists among science's greatest failures, like those who continued to work on Ptolemaic epicycles...[244] – *Lee Smolin*

All indications suggest that string theory, which for decades has tied up the brainpower of the most talented mathematical physicists, is one of the greatest wastes of intelligence in modern times. A disturbing monoculture prevails in the universities,[I] strengthening groupthink and leading to an arrogance of power. Today, they no longer even care about providing any experimental evidence for their

---

[I] Thus, more than ninety percent of the leading chairs in theoretical physics are occupied by string theorists, cf. Smolin (2006)

claims. This means that the empirical method as such, on the basis of which physics has been successful since Galileo Galilei,[245] has been abandoned. If one presupposes a minimum of self-reflection, it is hard not to interpret this as dishonesty. Certainly, the system is beyond repair.

Where all think alike, no one thinks very much. – *Walter Lippmann*

DO NOT INSULT THE PROPHET

When the sun of a culture stands low, even dwarfs cast long shadows. – *Karl Kraus*

With regard to the prominence of string theory, a special role belongs to the American mathematician Edward Witten, who for decades has been considered the intellectual leader of the movement. Awarded the highest mathematical distinction, the Fields Medal, Witten is regarded as a genius and, although he has not yet published a single genuinely physical result, is often portrayed as Einstein's successor. This is also due to Witten's position at Princeton, Einstein's last place of activity, which is often used as propaganda for string theory. A particularly blatant approach is adopted by the mathematician Brian Greene, who insinuates that string theory is in fact a continuation of Einstein's ideas. One of Einstein's collaborators at Princeton, the mathematician Theodor Kaluza, had worked on a five-dimensional theory of spacetime, but conceptually it had nothing whatsoever to do with strings.

The pure mathematician, however good, understands nothing at all about physics.[246] – *Werner Heisenberg*

Well, Witten may be mathematically brilliant. However, his cult status alone, which he cannot be unaware of, and the fact that for decades no palpable results have emerged from these collective efforts should actually lead him to seed his scientific entourage with a little doubt; should he have any sense of responsibility left. Apparently, we are dealing with the not-so-rare case of great talent being accompanied by a certain lack of common sense, which, of course, has nothing to do with being modest.[247] To this day, Witten retains his congregation with the preposterous narrative that string theory "predicted" gravity and has thus practically already achieved the union of gravitation and quantum theory.[248] The reader may well sense the exuberance of this statement, but its absurdity becomes clear as soon as one takes a look at the cosmological work of Paul Dirac.

QUANTUM LYRIC MEETS HARD NUMBERS

God created everything with number,
weight, and measure. – *Isaac Newton*

Testable physics generates quantitative assertions. The ratio of forces between electric and gravitational attraction of electron and proton, which form the hydrogen atom – i.e. the simplest quantum system – is $2.27 \cdot 10^{39}$, a huge number with almost 40 digits. Every theory that aspires to unify physics in the direction of a "quantum gravity" must necessarily calculate this number if it does not want to be left with empty platitudes.

When you cannot express it in
numbers, your knowledge is of a
meagre and unsatisfactory kind.
- *Lord Kelvin*

As mentioned in Chapter 6, Paul Dirac had related this number in 1938 with the size ratio of the universe to the proton, which also has about 40 digits. Moreover, the total number of particles in the universe is approximately 10^{80}, the square of the previous number. Admittedly, this fascinating coincidence never evolved into a complete theory. One may therefore doubt whether Dirac's approach will succeed or even question it in general. Indisputably, however, it is the only quantitative idea in the history of physics concerning that enigmatic number that appears in the hydrogen atom. Therefore, to talk about "quantum gravity" without addressing this issue, or at least pursuing another computational alternative, therefore simply demonstrates historical ignorance.[249]

Chapter 14
The Loss of Truthfulness
Arbitrary physics

> One cannot take away integrity in the search for evidence and honesty in declaring one's results and still have science. – *Harry Collins*

At the risk of sounding old-fashioned, science is not a business; it requires idealism. Science without honesty is not only bad science, it is no science at all. For science is defined as the search for truth. And today, this search is no longer in good shape.

We are not primarily referring to frauds such as those of Emil Rupp, who was exposed in the 1920s by experimentally validating an error in Einstein's calculations, or of Jan-Hendrik Schön, who was considered a shooting star in nanophysics until his forgeries were exposed in 2002. While these incidents are enlightening in terms of the sociology of science, they do not represent the essence of what fundamental physics suffers from. Loss of truthfulness rather means that there is no longer any serious search for knowledge at all.

> Scientists are usually too cautious and timid to risk telling outright lies … but they push the envelope of exaggeration, selectivity and distortion as far as possible.[250] – *Bruce G. Charlton*

Cheaters must fear discovery. In this respect, they even take an adequate risk that counterbalances the scientific fame they have gained. Those researchers who superficially obey the rules, but are not truly committed to the study of nature contribute more to the erosion of science, which constitutes a serious violation of scientific ethics. If one considers these ethics in a strict sense – which is what respect for the discoverers of the past four centuries demands – science is the pursuit of knowledge about the laws of nature, devoid of any other interests.[251]

TRUTH SEEKER AND FUNDRAISER

Hardly anything could be further from this ideal than today's scientific establishment. The number of studies that are not reproducible has become alarming[252] – which already makes it obvious that many do not take things very seriously. In addition, a great many of today's researchers are involved in projects they are not really convinced of themselves. They understand that these projects are neither of practical use nor do they lead to new insights. They do not believe in the promises they write in their own research proposals, and they are well aware of the embellished sections.[I] Moreover, they would never pay for their research with their own money. These unpleasant truths have been spelled out by Bruce G. Charlton[II], a knowledgeable expert on the contemporary scientific industry, in his book *Not Even Trying*. What the title means is: They do not even try to be truthful any longer.

[I] According to a recent survey, 'lies and exaggerations' have become routine in research proposals (Hossenfelder 2018, loc. 2545).

[II] Charlton was a long-standing editor of the journal *Medical Hypotheses*. Nothing suggests that the situation is any different in fundamental physics.

The pursuit of truth is more delightful
than its secure possession.
- *Gotthold Ephraim Lessing*

If we take genuine truth-seeking as a yardstick, much of today's institutionalized science is thoroughly dishonest. In fact, almost everyone is aware of this. By tacit agreement, however, much is tolerated as long as certain rules to which funding is tied are followed.

A moral grounding is indispensable for a scientist. There should actually be a "Hippocratic Oath" in pure science as soon as one chooses this path, which requires a strong sense of responsibility. At the same time, individual researchers should become as independent as possible from institutions and be committed only to the search for truth. This seems difficult to realize, all the more so in today's culture where prestige is increasingly defined by income.

Yet now I regard real science – the
kind of science I used to worship - as a
thing of the past. – *Bruce G. Charlton*

SHAMELESS BOASTING

To some – such as perhaps David Gross or Edward Witten – we might concede that they do regard themselves as truth-seekers in their self-congratulatory spinning around their own theories. The majority of modern celebrities, however, take their privileges for granted and would regard anything like a responsibility or even obligation to seek the truth as outlandish. When the abovementioned theorist Alan Guth touted his ambitious but completely spaced out speculation of "cosmic inflation" in 1979, his colleague Lenard Susskind remarked, "You know, the most astonishing fact is that we are even being paid for such things." Such statements elicit smiles and may sound pleasantly self-ironic, yet they reveal a grave lack of the seriousness that was part of the physicists' professional ethos at the beginning of the 20th century – even though they were certainly used to humor in their dealings with each other. But it would never have occurred to them to engage in banter about how absurd an idea sounds. To sound absurd has become a hallmark of modern physics, whenever it wants to appear creative.

It has become customary is to camouflage the absurdities by blithely addressing them, as Arkami-Hamed did, for example:[253] "So there's no experiment and you just sit around and talk about beauty and elegance and mathematical loveliness. And it sounds like sociological bullshit." And he gets away with such brashness. Once uttered flippantly, he has cleared himself of the thoroughly justified charge and is confidently lecturing on "the morality in fundamental physics."[254]

At times, Arkami-Hamed and other popular theorists such as Brian Greene frankly comment on the possibility that it may well be that everything they are dealing with is nonsense.[255] This is not as self-reflective as it sounds, for people typically grant this realist conjecture only to themselves, the experts, but not to anyone outside the *community*. Most importantly, such an attitude reveals a lack of the responsibility that any truth-seeking researcher should feel. These gentlemen do not care how much funding society provides for their privileges, in the good faith that they are doing something for the long-term progress of civilization. They do not care if billions of dollars are invested based on their half-baked speculations, tying up substantial technical and intellectual potential of humanity. Yeah, it could all be nonsense, sorry! They are not ashamed of anything.

> We need to build this collider because we can. – *Nima Arkami-Hamed, 2020*

SILENCE FOR MONEY

For a large number of theoretical physicists it has become perfectly natural to spend their entire professional lives dealing with things they basically believe are pure fictions. Nor is such dishonesty excused by the fact that some may find calculations in ten dimensions or in the first 10^{-35} seconds fascinating. There is nothing wrong in dealing with abstract mathematics unrelated to reality, if this was communicated and the public could decide to what extent it would finance such gimmicks instead of pursuing really fundamental questions of nature.

However, mathematical physicists, in particular, often corrupt themselves in their publications and proposals by referring to a potential application in physics, although they know perfectly well that this is just a fig leaf. For anyone with a basic knowledge of the philosophy of science, the absence of experiments and thus the lack of falsifiability is evidence that this is not proper science. That is why in recent years there have even been attempts to rewrite the theory of science based on Karl Popper in a way that would eventually allow dealing with such nonsense. Today, lectures are held on whether theories can be confirmed "non-empirically."[256] Conferences take place where people discuss in all seriousness how to get along without the burdensome requirement of evidence in the future: a rebirth of mythology[257] that still calls itself physics.

This obvious perversion of methodology has a detrimental effect beyond science. If effective demarcations against irrationality no longer exist even in science, the floodgates are opened to the erosion of modern society by any kind of ideology. The relinquishment of evidence is reminiscent of the large-scale scams in the economic and financial systems, which have often become legal thanks to lobbying. In his bestseller *The Black Swan*, Nassim Taleb describes this separation of the ethical and the legal as a characteristic of modernity. It seems that in the modern age, institutionalized research and the search for truth are just as far apart.[258]

We have managed to transfer religious belief into gullibility for whatever can masquerade as science. – *Nassim Taleb*

IF YOU CRITICIZE, YOU ARE A POPULIST

Supposedly objective, the scientist does not voice opinions outside his field of expertise, although privately he also considers many theories to be nonsense. The loss of truthfulness becomes very clear when one listens to the conversations of experimental physicists about theories such as superstrings or cosmic inflation.[I] Everyone knows that this is not serious science. Nonetheless, people avoid saying it in public or even in research proposals. After all, the bursting of the speculation bubble in theoretical physics would, among other things, wreak havoc on the ramshackle models that high-energy physicists use to justify their occupation. Much of the experience of *senior scientists* consists in

---

[I] Even an early proponent of inflation, Paul Steinhardt, now argues that it cannot be tested any longer "because of the large number of models." (Hossenfelder 2018, loc. 3433).

skirting these cliffs and mastering the misleading narratives that sell banal research activity as revolutionary insight. Ultimately, this has become an openly practiced, almost natural intellectual corruption.

> A vast, international activity with millions of employed workers and multiple billions of dollars of funding – is so thoroughly corrupt as to be unreformable.[259] – *Bruce G. Charlton*

Just as citizens lack the power, money, and influence to stand up against large-scale, legal corruption in business and politics, the academic institutions benefit from the public's lack of information about what researchers actually do. Apart from professional marketing by the press departments of the *big science* organizations, there is little that could provide guidance or even transparency to ordinary people. Just as the political elites insinuate that the public is too stupid to decide on certain things, top-level researchers protect themselves from unpleasant questions by arguing that non-experts are too ignorant to understand them.

> I think my soup is salty: may I not call it salty until I am allowed to cook it myself? – *Gotthold Ephraim Lessing*

Whoever raises doubts about whether the next billion-dollar accelerator makes sense,[260] is by definition incompetent. Non-string theorists are not qualified to offer an opinion on string theory. Under no circumstances can particle physics be criticized by anyone who has not been brainwashed for decades by its beliefs. Presumably, the time is not far off when dissent from mainstream cosmology will be labeled as a fake news and conspiracy...

Theoretical physics has gone astray. But it is hard to predict when a bubble will burst. Not only have people invested financially, but many have built their careers on it, which causes enormous psychological reluctance. It is not easy to admit that one has been beguiled by phantasmagoria all one's professional life. Much remains to be desired as far as honesty with oneself is concerned. The *Sunk Cost Fallacy* reliably prevents the insight that what has been invested so far is lost. The end of this misdirected research activity, which will come sooner or later, can therefore be compared less to a dip in the stock markets than to a collapse of the entire financial system.

It is hard to make somebody
understand something when his income
is based on not understanding it.
– *Upton Sinclair*

SCIENCE HAS ITS CORPORATE MEDIA

Just as today's political system is backed by the media, where certain topics or critical questions do not even surface, science can be assured of the loyalty of its journals. This does not refer primarily to specialized publications, which use peer review to protect themselves from radical criticism. What is more serious is the subservient tone of popular science magazines, which portray science in bright colors. In applied physics, this is often warranted; critical reporting on fundamental research, on the other hand, is practically unthinkable; even the most absurd theoretical fantasies are obediently recounted and nicely illustrated. Investigative journalism (which is having an increasingly difficult time anyway) simply does not exist in science. In addition to the well-known reasons for this – insecure working conditions and herd instinct – there is a lack of competence here. Even the few journalists trained as physicists cannot afford to tackle a specialized field without being accused of not having the necessary expertise. As so often, the fear of being considered a troublemaker outweighs the courage needed to be a whistleblower.

The *Physik Journal*, for example, a lobby sheet of the German Physical Society, focuses primarily on funding and, as a result, regards all research as great. But even the journals one pays for read like yellow press when it comes to topics like particle physics. Because no one can explain the subject itself anymore, they write sentimentalized stories about the researchers (preferably female ones), what they feel, what their dreams are, and so on. The illusion that stories about scientists are actually about science was even taken to the cinemas with the movie *Particle Fever*. It is full of grotesque misrepresentations, such as that the mass of the Higgs boson would make it possible to draw conclusions about "supersymmetry" or the "multiverse" or even give "clues" about a theory of everything.

> The Higgs doesn't take us any closer to a unified theory than climbing a tree would take me to the Moon.
> – John Horgan

The symbiotic relationship between media executives, journalists, and science managers is not one iota more honest than the *Trading Truth for Access* in politics. The experts allow for proximity in background conversations, while the journalist skips the sensitive topics. Any upright reporter who challenged this kind of research with real questions would soon be out of a job. Therefore, it is more convenient to be invited to guided tours and seminars by the lavishly equipped press departments – at CERN, for example. It is surely no coincidence that not a single critical question was asked during the major physics press conferences that have been covered by the global media in recent years. That does not even happen in politics.

The inevitable result is an injection of journalistic flavor into Big Science which is fundamentally in conflict with the scientific method. – *Alvin Weinberg*

LONG-GONE NOBLE

The Nobel Committee – or rather the Swedish Academy of Sciences – must also be considered part of the ecosystem that is called science today. Many epochal achievements in science have been awarded prizes from Alfred Nobel's heritage. Yet this institution, too, is only a mirror of the scientific community of its day. Since practically none of the committee members has the expertise[I] to assess the achievements proposed for the prize, they have to rely on external referees who, in turn, are firmly anchored in the respective research communities.[261]

There have been justified awards, such as for magnetic resonance imaging or the laser frequency comb, but also completely silly ones, such as for the discovery of "neutrino oscillations"[262] or "asymptotic freedom".[263] In order to get themselves talked about in Stockholm, in recent years it has become standard practice for research collaborations to organize press conferences with great fanfare at which they present their hitherto secret results in a sensationalist

---

[I] Admittedly, this was already the case with Einstein's theory of relativity, leading to the decision to award him the prize for the light quanta hypothesis in 1921.

manner. The scientific general public is taken by surprise and denied the opportunity for independent evaluation.[I]

> In matters of science, truth is the ultimate sensation. – *Oscar Wilde*

More in-depth investigations then often lead to open questions whose significance is no longer realized because the discovery has long since been "established".[264] This dubious practice has some parallels with international politics, where sanctions are imposed or military strikes are even carried out after claims have been made about issues that subsequently turn out to be far from clarified.

> Certainty is one of the cheap commodities, and it can be obtained instantaneously once the problem has been tackled in the right way.
> \- *Paul Feyerabend*

DOES EVIDENCE MEAN NO CONTRADICTORY EVIDENCE?

An example of particularly brazen behavior was exhibited in 2014 by the BICEP2 collaboration, which claimed to have found traces of gravitational waves shortly after the Big Bang in its cosmic microwave background data. I personally attended a lecture by a U.S. theorist who immediately heralded this as a major breakthrough. The claim collapsed only because grave technical mistakes in the data analysis were revealed.[265] The much greater absurdity, however, was to have extrapolated the microwave background[266] (when the universes was 400,000 years old)

[I] Prime examples were the Higgs boson in 2012, gravitational waves in 2016, or the 'picture' of a black hole in the M87 galaxy in 2019.

back to the first fractions of a second after the Big Bang. Physicists often complain about the charlatanism of homeopathic substances, not a single molecule of which is left after appropriate dilution, but this kind of argumentation is not one bit more credible.

It is entirely normal and unremarkable for ordinary 'scientists' to spend their entire professional life doing work they know in their hearts to be trivial or bogus.[267] – *Bruce G. Charlton*

The modelers of so-called cosmic inflation, in particular, are fond of postulating processes shortly after the Big Bang and claiming that there is evidence for them in the cosmic microwave background. If one were to illustrate this with an analogy, one might compare it with expecting to discover deep-sea creatures by observing the surface of the ocean and to classify them in zoological terms. For this, you first convince yourself of the existence of a fish by theoretical reflection and then call it "evidence", if the waves do not prove the opposite...

Those who can make you believe absurdities can make you commit atrocities. – *Voltaire*

This may seem preposterous, but it is in fact an argument that can be encountered in modern physics. It is also known as Hempel's paradox, after philosopher Carl Gustav Hempel. The statement "all ravens are black" is mathematically equivalent to "what is not black cannot be a raven." This is straightforward logic, but it becomes interesting when we look at the evidence for the assertion. The observation of a black raven certainly speaks in its favor; but so does the observation of a yellow shoe: after all, it is an object that is neither black, nor a raven. No rational person will take the latter "evidence" very seriously. In physics, however, such reasoning is widespread – even if a "raven" has never been seen and one does not know what "black" means.

Whoever does not lose his mind over
certain things has no mind to lose.
- *Gotthold Ephraim Lessing*

TELLTALE VOCABULARY

Besides these obvious distortions of the search for truth, there are many minor self-deceptions that researchers like to indulge in. The same kind of "seeing by not seeing" is the case when missing energy in particle detectors is interpreted as neutral particles, of which there are plenty. The evidence for what is called dark matter, which is the very absence of luminous signals, is ultimately of this type as well. Nevertheless, the journals are full of reports that enthuse about the "first direct" evidence of dark matter, albeit usually with the caveat "should the results hold true." [268] In many of today's popularizations of physics, this dishonest use of the subjunctive leaps to the eye.

A recent statistical study showed that in the publications of the past few decades the use of exaggerated and misleading expressions has increased more than twentyfold – expressions such as "unprecedented," "groundbreaking," "novel" and "promising."[269] One can hardly say any more about the decline of the scientific ethos.

Most scientists today are devoid of ideas, full
of fear, intent on producing some paltry result
so that they can add to the flood of inane
papers that now constitutes "scientific
progress" in many areas. – *Paul Feyerabend*

If researchers express themselves so disingenuously, it is hard to blame journalists: "If these calculations are correct," "if the smartest theorists are not mistaken," "if this theory really represents reality" are phrases that are used to introduce even the crudest nonsense in science programs.

Sadly, it sometimes seems that viewers are all too eager to indulge in such fantastical stories. The popularity of fantasies like wormholes, high-dimensional "branes" (a generalization of strings) or "multiverses" even among lay people is an amazing phenomenon. Some science fans prefer to be dazzled by some exotic idea rather than to listen to a clear but exhausting line of thought. If something is so obviously outlandish, at least people do not feel alone if they do not understand anything. Thus, hardly anyone is offended when basic physics is casually blended with science fiction and descends to the level of a film industry's idea generator. It is hardly possible to move further away from truthfulness.

BREAD AND GAMES FOR THE MIND

It keeps astonishing me how popular such speculations as extra dimensions, parallel worlds or superstrings, and even black holes, have become. Beyond the mere words, these things are difficult to explain in a general way. On the other hand, it is quite satisfying to have understood the Gaussian integral theorem or the general gas equation, even if this may sound relatively dry. But one should not be surprised about these distorted criteria – all this is part of a culture in which entertainment counts more than education.

All great minds have felt a state of *flow* (in the sense of the motivational researcher Mihály Csíkszentmihályi) when delving into the fundamental laws of nature. Einstein's joy over the derivation of the Mercury orbit by the general theory of relativity in 1915 was so great that he could not fall asleep his heart was pounding so hard.

By contrast, one often encounters modern physicists who refer to their work as "fun". The intention is probably to free fundamental research from the demand of direct usefulness. However, this is short-sighted. True basic research *is* useful, because today's civilization has emerged from it. Just think of Heinrich Hertz, who, out of scientific curiosity, wanted to prove the existence of electromagnetic waves. In the long run, such non-targeted research actually pays off. But since today's research culture (and probably other kinds of cultures too) lacks long-term thinking, its justification falls back on "fun". Science is about reality, however. Mach, Einstein, or Schrödinger would never have dared to describe their engagement with the laws of nature as fun. Instead, they considered it a responsibility with respect to creation and humankind.

Also in science, American culture favors what bestselling author Neil Postman called "infotainment." The popular TV series *The Big Bang Theory* entertains the public by means of theoretical physics. It portrays as nerds those who, in an unswerving belief in their theory, indulge in incomprehensible calculations, something that can be amusing. Alas, the appeal of satire dwindles to the extent that it matches reality. One wonders what would happen if the general public became aware that theoretical physicists are actually very similar to their actors. So far, however, the belief in parody seems to sustain the image of the entranced geniuses.

When serious public conversation becomes a form of baby-talk... then a nation finds itself at risk; culture-death is a clear possibility. – *Neil Postman*

The origin of this annoying state of affairs ultimately lies in the incomprehensible behavior of nature, which first came to light in the 1920s. At that time, people had to admit that the micro-world looked "crazy" in the sense of "puzzling". The failure of scientists to explain these observations in a clear, reasonable, and logical way has led, in an act of psychological escapism, to a situation where physicists now allow themselves "crazy" theories, this time, however, in the sense of "devoid of any relation to reality." Because common sense cannot be switched off completely, many physicists take refuge in an apologetic self-irony when dealing with ideas that are crazy in every respect.

ANY INTERPRETATIONS

It may occur that a young researcher is blindly enthusiastic about an idea. But if, as it happens, 500 (!) theoretical papers appear that "explain" an anomaly in the CERN data, which then turns out to be a statistical fluctuation,[I] then every physicist with a shred of professional self-reflection should admit that something is wrong in the system. Modern particle physics has not been falsifiable for quite some time.

I do not mind if you think slowly, but I do mind when you publish more quickly than you think. – *Wolfgang Pauli*

[I] Such as the 'diphoton anomaly' at the *Large Hadron Collider* 2016, see Hossenfelder (2018), loc. 1403.

Mind you, it is not only the number of these meaningless articles that is disconcerting, but the fact that every deviation from the theory observed in recent decades has led to similar knee-jerk reactions.[270] Many youngsters apparently idolize theorist Gordon Kane from Princeton, who in 2012 used the leaked measurements of the mass of the Higgs boson[I] to "predict" that very value in his string model.[271] Such people have lost all decency.

You can make a living from prophecy,
but not from telling the truth.
- Georg Christoph Lichtenberg

It is hard to believe that all these physicists want to explore nature in good faith. At best, their herd instinct helps them to displace the truth. It is only a small step from such dishonest attitude to cynicism, which justifies itself with the argument that all the others did it in such a way. Among the legions of physicists, quite a few have probably made this inner resignation already.[272] Cynicism is not only no science, but sabotage of mankind's enterprise to understand nature.

You can't lie with math. But it greatly
helps obfuscation.
- Sabine Hossenfelder

DANCE AS LONG AS THE MUSIC WILL PLAY

All in all, physicists have developed a system that keeps itself alive by detaching from observation, by abandoning comprehensible mechanisms and clean mathematics, by

---

[I] This is not to say that the Higgs boson is a useful concept. Details of the history of its discovery can be found in Unzicker (2013) and Comay (2014).

postulating arbitrary concepts, and by weaving a "theoretical explanation" for every thinkable phenomenon, which amounts to nothing other than the fitting of fantasy products to measuring values.

However, anyone who observes these activities with open eyes and a degree of perspective will sooner or later become aware of this fact. Yet honest self-reflection is not a virtue that thrives in today's institutionalized research environment. These institutions woo funding with an often-repeated marketing message that sounds roughly like this:

It's easier to *fool people than* to *convince them* that *they have been fooled.* – *Mark Twain*

We are already far advanced in our understanding of nature; admittedly the standard models are not perfect, but the next experiments will certainly provide clues to a more fundamental theory. In short: Give us more money and we will find something that we will somehow interpret and celebrate as a confirmation of our standard models. Or we might find nothing at all, which would be even more awesome, because it points to physics "beyond" these models.[I] That narrative is thoroughly mendacious. It should go something like: We have no idea; we are aware that we are not predicting anything, and we feel quite clearly that we are at a dead end. But before you pull the plug on our funding, we will throw a lot of dust in your eyes. This is where fundamental research stands.

[I] This is not a parody. In 2013, the Nobel Prize was awarded for a confirmation of the Standard Model; in 2015, the discovery of a deviation from it was considered worthy of the prize.

Chapter 15

Postmodern Science

Specialization to the Point of Crumbling

> Don't get involved in partial problems, but always take flight to where there is a free view over the whole single great problem, even if this view is still not a clear one. – *Ludwig Wittgenstein*

It appears that institutionalized basic science is facing a crisis. So why does this well-organized endeavor, which is rooted in human curiosity, fail so blatantly in the long run? Science shapes the development of our civilization, influencing every aspect of human interaction. We are not talking about technology only, but also about consequences for law, ethics, and society.

The amount of accessible knowledge has increased dramatically since the beginning of modern science in the early 17th century. Nevertheless, 100 years ago that knowledge could still be regarded as a whole. The science historian Derek de Solla Price[273] recounts that he has read all the volumes of the *Philosophical Transactions of the Royal Society of London* published from 1662 to 1930, line by line. A comparable effort would be unthinkable today; the journals of just one specialized field occupy entire libraries. It is becoming increasingly difficult for individuals to grasp the knowledge of mankind in a compact and accessible form.

GROWTH, BUT NOT IN CREATIVITY

The accumulation of knowledge generated new fields of research and application. This required more and more scientists, which was a problem in itself. Real progress was made by single, creative geniuses who, even with a growing world population, only appeared in limited numbers.[274] The precondition for this is, above all, a proper intellectual environment. As there was no longer a sufficient "supply" of these geniuses, science began recruiting clever, but less creative people, who devoted themselves to the abundance of scientific-technical tasks. These are well-trained researchers who are highly skilled in their fields, but not the kind of people who trigger crucial breakthroughs.

People who make great discoveries somehow manage to free themselves from conventional thinking.
– *Anthony Leggett*

The actual system in the U.S. really discourages people who are truly original thinkers, which often goes with slow maturation at the technical level.[275] – *Lee Smolin*

Using the terminology of philosopher Thomas Kuhn, these would be called "normal scientists," in contrast to the visionaries who think outside the box and who have unleashed scientific revolutions by their uncompromising search for the truth.

Certainly, there are creative minds in science still today who are changing the world with their discoveries, especially in experimental physics. However, this is accomplished rather by individuals or small, motivated groups, mostly in an uncompetitive climate of academic freedom.[276] Respectable, yet less genuine science is the chase for world records in precision measurements, which already ties up more manpower and resources. By contrast, the least benefit is gained by ambitious large-scale projects that seek to verify theoretical wishful thinking.

MASS SPORT SCIENCE

Everyone agrees that genius is entirely opposed to the spirit of imitation.
- *Immanuel Kant*

Science as a mass phenomenon has only existed in U.S. culture since the postwar period when the *big science* collaborations were established. What resulted was a professionalization and specialization of science. It is somewhat curious that the term "professional" has any positive meaning at all in research. After all, the list of "amateurs" who made decisive breakthroughs is quite long: Michael Faraday, Andre-Marie Ampère, Wilhelm Herschel, Johann Jakob Balmer, Albert Einstein, to name just a few. In addition, many of the founding fathers, such as Niels Bohr, Paul Dirac, Werner Heisenberg, Erwin Schrödinger, and Wolfgang Pauli, were either still students at the time of their most significant discoveries or were working as professors outside their main duties. Certainly, they did not apply for research grants, raise third-party funds, or found societies. The search for truth has nothing to do with money and prestige. It is a profound misconception that the progress of fundamental science is limited by the resources available to it. Institutionalized science is just as much a contradiction as paid idealism.

In fact it may be impossible to get a job, or get tenure, or promotion - except by dumping idealism and scientific ambition and embracing low-risk careerism.[277] – *Bruce G. Charlton*

This is another inherent contradiction researchers struggle with. Being a scientist is a vocation, much more than a profession, and certainly not just a job. Yet science has transformed from a meaning of life to a means of livelihood. A large part of those working in research today have swapped scientific aspiration for career ambition. The primary motivation of a scientist, the search for the truth, has long since ceased to be a qualification. Rather, to be guided by scientific ethics in one's actions beyond what is expected by the working environment has become a hindrance. In "modern" science, which has spread all over the world as a consequence of the U.S. culture, such basic values have long been missing.

Real science is … incompatible with the idea of … a career.[278] – *Bruce G. Charlton*

PROFESSIONALIZATION OR DE-IDEALIZATION?

It is not enough to teach someone a specific subject. That makes him a kind of useful machine, but not a mature personality.[279] – *Albert Einstein*

Amateurs are usually all-rounders, while professionals like to regard themselves as specialists. In fact, a certain degree of specialization is inevitable and beneficial for the thorough study of a subject. It is worth remembering, however, that most natural scientists at the beginning of the 20th century identified as generalists. They penetrated deeply into particular fields, but considered it a must to have an overview of physics as a whole. Prominent examples are Einstein, Schrödinger, and Dirac, who had worked on both quantum physics and cosmology. As natural scientists, they were simply too curious to leave out other fields.

By contrast, today's physics is specialized to a degree that technical terms alone make it difficult to communicate across ever-narrower boundaries between the fields. Some people bemoan the difficulties by calling for a new Einstein to emerge who would be able to have an overall view of and understand physics as a whole. Among eight billion people it would actually be surprising if there was not someone with comparable abilities. It is more plausible, however, that physics itself has become so fragmented that no one can in fact view it as a big picture any longer.

Early specialization guided by immediate usefulness ... kills the mind.[280] – *Albert Einstein*

BLINKERS FOR EVERYONE

This increasingly refined micro-specialization has certainly exceeded the optimal level today – just look at some exotic university programs. Due to the corresponding fragmentation, no one can any longer check the respective results across disciplines. Science has increasingly become a hodgepodge of isolated and hard-to-refute narrow-track expertise. A neutrino physicist does not understand enough about galaxy dynamics to have a say in that field; and neither would the arguments of a quasar expert be taken seriously by accelerator physicists. As a result, the branches of fundamental physics communicate only by superficial half-knowledge. People have to rely on opinions of authorities, who in turn rehash the group consensus and must do so in order to be recognized as authorities.

The picture of the nerd – a highly intelligent specialist who loves to talk in technical jargon – fits perfectly with the type of physicist in the dominant U.S. research culture from the 1980s onward. In his book *The Trouble with Physics*, Lee Smolin describes the herd mentality physicists displayed when it came to the latest hot topic.[281] The typical nerd is therefore by no means a strong character, but rather part of a *community* that provides him with ideas and motivation. His ambition is not to discover anything fundamental, but just to be the first to find out something.

Ambition is the death of thinking.
- *Ludwig Wittgenstein*

Here again, one can see the contrast to physicists of earlier times, whose way of doing things was diametrically opposed to today's approach: Johannes Kepler was already a loner as a child; Newton spent years living in isolation on a country estate; Einstein had good friends but scientifically remained a solitary fighter; Dirac's reserve bordered on autism; and Schrödinger let his work be inspired mainly by his lovers.[I] In any event, they all followed their own paths and had just one thing in common: the desire to reveal the secrets of nature by putting all their energy and dedication into it.

Can we even compare this with today? People claim that science has changed because "all the easy problems have already been solved" and modern physics is just hard. The individual, they say, is overchallenged with such tasks. As

---

[I] "Schrödinger's wife had long since given up objecting to Erwin's love affairs in any form, especially when Weyl was around. The ordinary standards of sexual morality did not apply here." (Biographer Walter Moore).

far as theory is concerned, there is not a shred of evidence to support this view. Rather, it ignores the achievements of those individuals, usually out of ignorance about the historical facts. For the scientific revolutions in Europe were almost exclusively initiated by lone visionary thinkers.

> New scientific ideas never emerge from any kind of institution, but from the mind of the individual researcher.
>
> - *Max Planck*

It is much more likely that the stalemate in theoretical physics is due to the collectivist nature of the efforts of recent decades, since creative individuals can hardly get their way in today's scientific environment. Another key factor is the ailing publication system, which is based primarily on journal publishers' profits. For the rest, it is an optimal breeding ground for uniformity of opinion.

IDEALISM UNDER SCRUTINY

> If anyone had submitted the ideas of those geniuses who were founding fathers of modern science to committees of specialists, there can be no doubt that they would have found them aberrant and would have discarded them precisely because of their originality and profundity.
>
> - *Louis Victor de Broglie*

Again, this shows a major discontinuity between the European and the American research traditions. Formerly, the editor of a journal decided whether to publish a submitted article based on his own competent judgment. This is how Max Planck in 1905 let an article "On the Electrodynamics of Moving Bodies" by an unknown patent clerk named Albert Einstein be printed in the renowned *Annalen der Physik*.

After the war, the formal *peer review* process, in which manuscripts must first be reviewed by colleagues before publication, became increasingly common. A positive aspect of peer review is that it allows expert feedback to be given. On the other hand, it became clear early on that revolutionary ideas that contradict the established wisdom of the field are often suppressed by peer review. For example, the landmark discovery of the citric acid cycle was rejected by the journal *Nature* as early as 1953. Astrophysicist Thomas Gold was not even allowed to submit a conference paper on his interpretation of pulsars[I] as neutron stars, which later would prove to be groundbreaking research. By the same token, countless low-quality papers and even computer-generated nonsense went unimpeded through the peer-review process. All in all, peer review marks a departure from the individual in the European tradition to the big collective research projects in America, in which the opinion of the *community* is the norm for what should be published.[II]

You could write the entire history of science in the last 50 years in terms of papers rejected by *Science* or *Nature*.
- Paul Lauterbur, 2003 Nobel Laureate

[I] Pulsars, discovered in 1967, emit an extremely regular radio signal through rapid rotation. Neutron stars are practically giant atomic nuclei with a diameter of about 20 km, although they can be heavier than the sun.

[II] Even the Internet provided only a brief heyday for scientific freedom of expression without peer review. The platform arXiv.org exercised a liberal policy for a long time, which has since given way to a rigid 'moderation' system. Even Nobel laureates such as Brian Josephson, for example, have been put on a no-publishing blacklist.

Only when there is sufficient consensus among researchers, the thinking goes, will it be worth allocating funds to projects. Such huge projects have traditionally been at the core of scientific activity in the U.S., which is increasingly institutionalized. Meanwhile, peer review has done much to transform science into bureaucracy, where creative ideas get buried in groupthink.

What had been the responsibility of an individual assessment, became the anonymity of an expert report; ultimately, we are dealing with nothing other than an opinion, prone to conflicts of interest or even abuse of power.

This is because reviewers are rarely held accountable for negligent or malicious assessments, and those being reviewed are largely without rights.[I] Above all, peer review lacks the corrective mechanism at large: Technical details get improved, but papers that question the foundations of the field as a whole are practically impossible to publish once the only peer reviewers available are those who have built their careers in that very field.

> The academic career puts a young person in the predicament of producing scholarly papers in impressive numbers - a seduction to shallowness that only strong characters manage to resist. – *Albert Einstein*

A poison for genuine science is the rampant use of evaluation and ranking, in line with an economistic mindset in which research must produce quantifiable results. This creates false incentives that have nothing to do with real science: Inflating an idea into multiple publications, citation

---

[I] Thomas Gold's idea of a scientific arbitration tribunal never caught on.

cartels, putting names on the author list that have not contributed to the work, and so on. Evaluation attempts to introduce what is believed to be healthy competition, but this only accelerates the departure from scientific ethics. It is easy to count articles, but progress cannot be measured so easily.

SCIENCE TURNS INTO BUREAUCRACY

It is bureaucracy, not competence, that is growing incessantly in scientific institutions. According to a survey published in *Nature*[282], less than half of actual working time is devoted to real research, while much time is spent on proposals and reports, which, by the way, are not paragons of truthfulness. Everyone knows that they are often as carefully worded as they are deliberately misleading, while the claims are only checked for internal consistency. At the root of all this is the erroneous belief that science can be conducted by allocating money and personnel according to a certain scheme. Yet, in day-to-day research, it is often only an elaborate facade of science that is built.

We act as if real science can necessarily be formalised, mechanised and made a process of mass production. And we don't even attempt to check whether this is true.[283] *–Bruce G. Charlton*

Likewise, history has shown that results cannot be forced even by the deployment of all possible resources. The quest for research funds, equipment, publications, recognition, and prizes, as well as the managing of institutions for this purpose, has nothing to do with pure science. Ultimately, everyone involved is aware of this as well. Real sensations were often a byproduct of untargeted activities, while many resources have been wasted to support theoretical fantasies.

Since no marketable products have to be produced by research, misguided developments are no longer corrected from outside. It is enough for the communities to celebrate their own achievements, even if no one else would buy this nonsense. In big institutions, the individual assumes practically no responsibility for his actions any longer; the large community serves only as a reassurance that what everyone is doing cannot be wrong.

ENTREPRENEURSHIP? NOT EVEN THAT

In his book *Skin in the Game*, Nassim Taleb argues that theorists, like entrepreneurs, should take an appropriate risk for their activities: Research should be unpaid, and success can be rewarded through prizes. He calls this a "deprostitutionalization" of science.

Taleb is also a fierce critic of academia, which he views as mainly generating bureaucracy, whilst placing obstacles in the way of creative minds. This entrepreneurial view is certainly something American, yet it would benefit many of today's research institutions – even if basic research seems difficult to organize in such a system. In fact, with gigantomania from America as well as bureaucracy from Europe, these institutions have today adopted the bad from both cultures. Much of science has become a response to funding opportunities provided by politicians, captive to the prevailing science fashions, who decide on money flow.

Science, and physics in particular, still generates exciting discoveries; just think of lasers, computers, nanotechnology, or quantum cryptography. These accomplishments serve as welcome cover for the sick areas of theory, while adorning physics as a whole and keeping its public prestige high. Do not forget, however, that virtually all of these applications date back to fundamental physics discovered by 1930.

The only exception is nuclear fission, which apparently works in reactors and bombs. However, there is nothing useful or, heaven be thanked, dangerous that is based on the discovery of neutrinos, quarks, W, Z, and Higgs bosons, let alone dark matter or dark energy. This is by no means to say that basic research should be governed by the dictates of usefulness. However, the absence of any useful application after such a long time indicates that many modern "discoveries" are figments of wishful thinking rather than real phenomena.

The reason we need science at all is because of the efficiency gains it brings as a long-term consequence. Applied physics benefits from the follow-up of the fundamental discoveries of the first half of the 20th century. In theoretical physics, on the other hand, it takes a very long time for the lack of productivity to become apparent; possibly longer than the current institutions are likely to exist.

POLITICAL CORRECTNESS – TIPPING SCIENCE OVER THE EDGE

In any event, instead of delivering results, these institutions are embarking on pointless navel-gazing. At CERN, a researcher was suspended[284] because, contrary to feminist doctrine, he argued in a workshop on high-energy physics and gender that it is not women but men who are discriminated against in physics. Martín López-Corredoira, a respected astronomer, calls what one must at least regard as an intrusion of ideology into science an "ideological witch hunt."[285]

As soon as an institution reaches a certain size, with the corresponding social dynamics, code-of-conduct programs and woke ideology are implemented. Even Google has not been spared this.[286] MIT recently cancelled a public lecture by a renowned climate researcher because he had made critical comments elsewhere about the *Diversity, Equity,* and *Inclusion* program.[287] One can think whatever one wants about these political issues, but it is clear that they have nothing to do with science. In addition to the institutions involved in what is supposed to be fundamental science, American universities have been feeling this crisis for a long time. This leads to an additional erosion of scientific standards.

Insanity is something rare in individuals. But in groups, parties, nations and epochs, it is the rule.
- Friedrich Nietzsche

In the end, this is also a consequence of a tradition of thought that manages everyday research but lacks epistemological reflection. Debates such as those between Ernst Mach and Max Planck about scientific method have long since disappeared. Last but not least, modern physics has neglected the demand for evidence to such an extent that there is no longer a criterion to separate the wheat from the chaff. As a result, educational institutions are open to whatever ideology is at hand, and the observed craziness is only a symptom of this vulnerability.

The organization of universities, which in the U.S. have to compete for funding, has in any event made them susceptible to misguided social trends. While quota rules, gender speak and all kinds of political correctness first infected the humanities[I], they are gradually entering the natural sciences as well.[288] What is probably an irreversible development involves not only abandoning the principle of appointing the most qualified[289] among students and professors, but also a climate in which any irrelevant triviality becomes scandalized, usually amplified by a digital mob on Twitter. The bad habit of forced public apologies is reminiscent of an early form of totalitarianism.

Schrödinger, who fathered three children out of wedlock during his time in Ireland,[II] as well as Einstein, would have faced suspension, if not worse, at any of today's American universities.[III] Of course, this tells us more about the current state of universities than it does about Schrödinger and Einstein. At today's universities, every decision about personnel or content is overlaid with a debate about what might be interpreted as racist, sexist, homophobic, or transphobic. Any thoughtful questioning of mainstream opinion, by contrast, is branded a conspiracy theory and fought with moralizing outrage instead of factual arguments.

[I] For example, see the *Grievance Studies*, which have demonstrated the lack of any quality standards. The book by Pluckrose and Lindsay (2020) is also recommended.

[II] Times are changing quickly. Meanwhile, the woke outrage machine, on the basis of meager suspicions, has erased Schrödinger from the name of a lecture hall in Trinity College in Dublin.

[III] Here, too, the situation has turned upside down over time. While in 1911 the French press scandalized Marie Curie's affair with the married Paul Langevin, Curie was later enthusiastically welcomed in the USA. At that time, the new continent was still more tolerant than the 'old' one.

BEGINNING ORWELLIZATION

It is remarkable that universities, which for centuries were centers of freedom, intelligence and open-mindedness, have turned into hubs of ideological repression. Nassim Taleb thus refers to them as "nuthouses."[I] This situation may still be barely noticeable in some faculties, such as mathematics, and in Europe such developments usually come with some delay. But sooner or later they will undermine the quality of university education everywhere.

Fewer and fewer students will be willing to go into debt for low-quality, overpriced, and politically indoctrinated study programs, and this will dry up the universities financially.

On the other hand, the private sector will soon take care of the education and selection of its workforce itself. Google is already planning to certify qualified applicants with degrees that can be earned in a few months at a fraction of the exorbitant tuition fees.[290] In fact, the often dull university education[291] is already putting off talented and motivated people.[II] Whether Google, in particular, is likely to offer curious scientists a valuable perspective is another question.

Institutionalized basic research in physics no longer produces anything that would be useful and has thus inevitably become dishonest. Generally speaking, it no longer contributes to improving the efficiency of societies, which

---

[I] In light of disciplines such as 'ethnomathematics' or the discussion of 'basic colonial assumptions in Western science', this term certainly begs to be used.

[II] Elon Musk quit his master in applied physics two days into his studies. He commented that College was "basically for fun but not for learning."

in the long run must lead to a collapse of the scientific enterprise. Furthermore, over time, it will no longer be possible to conceal how far theoretical concepts have become divorced from reality. At some point, people will ask what benefit these activities provide besides sustaining those involved. As a result, these institutions will lose the very reason for their existence, simply because no truly fundamental research has been done for a long time. Yet the question is whether this will be part of an even more general instability affecting modern civilization.

Chapter 16

A Looming Crisis

The End of the American Age

I have pointed out in the previous chapters how fundamental physics has gradually passed its peak and entered a phase of degeneration. The roots of this phenomenon lie in a superficial school of thought that spread from America some 100 years ago and is now usually called "Western" culture. Sooner or later, this kind of theoretical physics will collapse once the public becomes aware of how counterproductive these activities really are.

However, the same non-sustainable culture of thinking has further implications for our civilization. As indicated in the first chapters, these implications could turn out to be so dramatic that the state of science would seem like a luxury problem by comparison. In fact, many systems are in danger of collapsing, and the prevailing mutual dependencies make predictions difficult. The only thing that is certain is that Western culture is not in a healthy state...

> The collapse of science is linked with the collapse of modernity - both as cause and as consequence.
> – *Bruce G. Charlton*

We have already mentioned the fragility of universities and of the education system as a whole. In addition, in a digitalized world, knowledge can practically no longer be kept secret or commercialized as such. Leaving aside specialized training, for example with the use of equipment, this will remove the financial basis of the educational institutions. However, one could also argue that the commercialization of knowledge, which stems from a misguided mentality, is now reaching its natural end. Ultimately, Europe's state-funded schools will not survive this crisis either. Moreover, in many subjects, it is becoming increasingly difficult for graduates to outperform artificial intelligence.

FALLING SHORT OF MAKING IT GREAT AGAIN

In Russia, the people are dumbed
down by the party; in the United States,
by television.
- Friedrich Dürrenmatt (1921-1990)

Education has always been expensive in the U.S., which has led to a lack of balance between the social classes. In recent decades, this gap has widened dramatically as the middle class has become impoverished, leading to an intensified social disparity. The days when climbing the social ladder could be achieved through hard work are long gone. Today, no one goes from rags to riches; the American dream has come to an end.[I] The America of the 1960s, where one could have a house, a car, a family, and prosperity through normal work has gone. The tents of the homeless line the cities of the West Coast, and drug abuse, spreading lifestyle diseases, rampant crime, and shootings have long since become the norm. In such a heated atmosphere, even a civil war along racial lines can no longer be ruled out in the long term.

[I] This is how Noam Chomsky describes the situation in his book *Requiem for the American Dream.*

The missing tradition of regulatory governance ultimately leads to a widening gap between the vast majority in economic decline and the approximately 300,000 super-rich.[292] Of course, this is also the result of a lax monetary policy that allows the big players to make arbitrary profits on speculation, while the risk is being hedged by the taxpayer. The money supply put into circulation has grown exponentially in recent years, as has national debt in the West. Since a turn to high interest rate policies would drive into bankruptcy not only the state, but also the majority of businesses, in the long run an inflationary overheating of the system is inevitable. In this respect, America, but also Europe, Japan, and Great Britain, have been bankrupt for a long time already.[293]

By provoking the war in Ukraine,[294] the U.S. has recently managed to drive Europe into economic suicide, the latter now probably the first to go over the cliff. However, the prospects for the entire West remain dire. This dead end is the result of a mentality that focuses on quarterly figures while lacking any long-term planning. Needless to say, this cannot be blamed on America alone; by now it has become a global failure of the system. What remains to be seen is whether a factual default will happen first in the U.S. or in Europe, whether there will be some other form of monetary reset, or whether financial capitalism will collapse altogether in its current form. Many people perceive this ominous situation more directly than the crisis of science. Yet the level of instability is similar.

PETRODOLLAR, ISLAM AND OIL WARS

The U.S. dollar has been able to survive only because it was used globally for the payment of crude oil. But this status as the world's reserve currency will not last forever, if only because the era of fossil fuels will end sooner or later. An increasing number of countries are escaping from U.S. monetary dominance by no longer settling their transactions in dollars.[295] A decade ago, Libya was punished for this with a *regime change*, but support for such wars of intervention is waning. Although the U.S. stirs up tensions with virtually all countries outside its empire, it can hardly afford to attack Iran, for example, which had dared to sell its oil without using dollars.

Oil price too low: U.S. cancels all wars until 2022. – *Der Postillon (satirical news site), 2020*

The intervention wars of the USA carried out in the Muslim countries of the Middle East had another disastrous consequence. In the best interests of civilization, it would be quite desirable to stand up against the intolerant religious ideologies that are widespread in these countries and to strive for an enlightened world order that is based on science in the true sense of the word. After all, the merits of the Occident do not lie about the fact that it was originally shaped by Christianity, but rather about that it developed a secular humanistic tradition.

However, this painful conflict with the church was fought in Europe a long time ago and has not been ingrained in the American mind to this day. On the contrary, American exceptionalism still exhibits religious overtones. How else can it be explained that during the 2003 invasion of Iraq, U.S. generals blathered about their God being the greater?[296] With the pointless destructions caused by these wars, the Western empire has forfeited any credibility to lead a sustainable and peaceful progress of civilization.

> The crimes of the United States have been systematic, constant, vicious, remorseless, but very few people have actually talked about them. You have to hand it to America. It has exercised a quite clinical manipulation of power worldwide while masquerading as a force for universal good. It's a brilliant, even witty, highly successful act of hypnosis. – *Harold Pinter, Nobel lecture 2005*

These wars, which in reality are waged for the profits of the oil, gas and arms industries, even with the propaganda campaigns of the U.S. media,[I] can no longer be presented as missions for democracy and human rights, especially when coffins are flown in from all over the world. At the same time, the Afghanistan war, justified by a multi-year pack of lies at a cost of more than two trillion dollars, was indeed a large-scale subsidy to the arms industry, yet certainly not a particularly efficient one.

Thus, leaving aside the rottenness of U.S. leadership – whoever that is – through illegal wars,[297] human rights abuses, torture, and drone killings, the question remains as to what long-term economic benefit the U.S. military empire is supposed to provide, especially since its excessive "defense" spending of over $700 billion – more than that of the next 10 countries combined[298] – is a major reason for its staggering national debt.

[I] Examples are the widely known incubator lie in 1990, the 1964 Gulf of Tonkin incident, or the Iraq war in 2003, which was demonstrably launched under false pretext. Extrapolations into the present are obvious.

Admittedly, this strategy of tension has had a major success with the new outbreak of hostilities in Europe. The U.S.-directed regime in Kiev seems ready to fight against Russia to the last Ukrainian, and European countries have even outpaced the U.S. in waging economic war that inflicts damage on themselves, while the U.S. energy and arms industry stocks are skyrocketing (needless to say, there is no benefit for Joe Sixpack). While it is physics that should remember the European values of one century ago, it seems that only the political leaders here are regressing into a stupidity that was widespread in 1914. In any case, one should not underestimate what is left of the power of the U.S. empire.

The naval, air, and ground forces, including the private armies that operate in three out of four countries of the world,[299] as well as the 800 military bases distributed around the globe, still provide a power and blackmail potential that can be used for economic purposes, also with the "allied" NATO countries. But it is anything but reassuring that the military, of all institutions, is the one that stabilizes the empire. Nations such as Russia, China, and India, which are virtually unassailable (at least directly) without the risk of a general apocalypse, have become independent from the actions of the United States. One wonders what would happen if the SCO,[I] for example, decided to turn the global chessboard around by imposing sanctions on the U.S.

[I] The Shanghai Cooperation Organization (SCO) is already far more important economically than the European Union.

The figures on the growth of American exports during the war baffled me. They were a real revelation to me. These numbers not only determined American intervention in the war, but also the pivotal role of the United States in the postwar world. – *Leon Trotsky*

THE OVERSTRETCHED EMPIRE

Although Russia currently leads technology in the area of hypersonic missiles,[300] the U.S. has of course little to fear militarily as long as it does not start a World War III itself.[I] The technological superiority of the CIA and NSA, which are capable of spying on and controlling the entire world, also supports the empire in such a way that an immediate meltdown is not to be expected. In the long run, however, one can expect it to collapse from within. Europe would therefore be well advised to become sufficiently independent so that it is not dragged into the abyss, even though that seems to be what is happening. In any case, the long-term dangers have not been averted, for the (voluntary, of course) recolonization of Europe in the superficial Western tradition of thought is continuing.

China, and Asia in general, with its immense, increasingly well-educated population, has in many respects surpassed Western civilization. It is buzzing with economic power and determinedly preparing for the collapse of Western currencies by buying up the world's infrastructure with

---

[I] Paul Craig Roberts (www.paulcraigroberts.com), former secretary in the Reagan administration, argues that the U.S. does not fear enough that possibility.

its dollar reserves. In this respect, their thinking is more long-term. Yet it is more perfection than innovation that is coming from these Eastern competitors. Chinese science will not flourish by building the pointless particle accelerators that the West can no longer afford.[301] In Eastern cultures, the individual has a lower status than in the West, which may be an advantage for the organization of societies, especially when thinking about living conditions. However, the accompanying lack of individual freedom, combined with a predominance of collective opinions, has also caused scientific breakthroughs by individuals against the majority opinion to take place mainly in the West. Even today, most creative innovators and researchers would hardly choose to settle in Beijing in order to realize their visions.

Being aware of these cultural differences in no way means joining in the anti-Chinese propaganda of the Western press. Arguably, the leadership in Beijing serves the interests of its own people more sincerely than the often corrupt elites of the West, which often construct only a facade of democracy that is pretty well controlled by the narratives of their media.[302] Officials like to bemoan the lack of freedom in Russia and China, although the restrictions at least help to maintain political stability.[I] The increasing suppression of free speech in Western societies, on the other hand, is at times reminiscent of a mass hysteria.

[I] Washington particularly appreciates the freedom to organize coup d'états all over the world. The blog https://caityjohnstone.medium.com provides insightful information on this.

For the West is now no more than a caricature of the culture of that liberty on which its civilizational success was based. If one thinks about the deteriorating climate of intellectual freedom, *cancel culture*, and censorship, and considers today's dissidents and political prisoners, such as Edward Snowden, Julian Assange, and Craig Murray, the Europeans Kant, Schiller, and Voltaire would be turning over in their graves, along with the Americans Lincoln, Jefferson, and Washington.

A POST-AMERICAN PERSPECTIVE

In his bestseller *The Clash of Civilizations*, Samuel Huntington provided many economic facts to prove that Western culture is in relative decline. One might speculate about which power might be the next to take over global supremacy. Nevertheless, I do not think this adequate in terms of the present state of the world. With all due caution about thinking the present is special, there are objective features that distinguish the current era from earlier epochs. Wars among the nuclear powers are no longer winnable; they are even more *lose-lose* than before. Yet, through madness or accident, the whole of humanity can be put in jeopardy at any time.

> I am not sure with what weapons World War III will be fought, but World War IV will be fought with sticks and stones. – *Albert Einstein*

Technology, more than anything else, determines the circumstances of living. Globalization has revolutionized the transport of goods and services; above all, the unhindered flow of information is changing the world. Thus, nationalities and the corresponding cultures of thought are not only merging through migration, but there are groups emerging worldwide that feel more connected than by mere nationality. Consequently, future trends in the history of culture can no longer be discussed from the perspective of national thinking traditions, as it was done at the beginning of this book. Meanwhile, all over the world, and of course also in America, there are researchers working and networking following a natural philosophical tradition,[I] while the culture of large-scale research institutions remains predominant.

It is interesting to ponder on the idea that the entire development described here was probably predetermined since the discovery of electricity, along with its waves and microelectronics. In any case, it is necessary to think in different terms than the centuries-old structure of nation states suggests.

Therefore, one can hardly expect that the U.S. dominance, which lasted for about 100 years, will in future be replaced by a corresponding dominance of, say, China. It is more likely that global history will enter a phase in which affiliation with an empire will play a smaller role, or that the era of national empires will come to an end altogether.

In fact, the influence of some internet corporations such as Google, Amazon, Facebook, Apple, Microsoft, and Twitter, which often brazenly exploit their monopoly positions, already exceeds the power of most nation states. The fact that the headquarters of these leaders are all located in America at least indicates that considerable innovation potential is still there and that, in this sector at least, the decline of the U.S. is still barely noticeable. Key structures of the internet as a whole, such as ICANN, are also firmly in

---

[I] For example, the *John Chappell Natural Philosophy Society*.

the hands of the USA. Although the above corporations still have to bow to state power, for instance via the decisions of the FISA secret court,[I] they do constitute a potential adversary to it. For the moment, rather the negative aspects are visible, because YouTube, Facebook, and Twitter, incidentally also Wikipedia, are heavily involved in the control, manipulation, and censorship of public opinion, which is on the rise throughout the Western hemisphere. As alarming as this is, this does not have to remain the case forever.

The value of these corporations consists primarily in intangible assets, software, and data. Therefore, in principle, they can settle anywhere on the globe, not to mention the fact that potential competitors can emerge just as quickly as they have.

NETWORKED NEW WORLD

At the same time, the internet players themselves are paving the way for the world to become increasingly interconnected, thus rendering nation states less important. Translation programs that facilitate communication among different cultures are certainly not unimportant. Along the way, internet corporations will dominate the education sector and make universities obsolete. All that is necessary is developing automated exams that can be scaled without much cost. Free access to education and the knowledge of humankind is one of the greatest visions the internet offers. Potentially, it will lead to a blossoming of talents that in earlier times would never have had the opportunity to develop their abilities. One might be optimistic

---

[I] It became known to a broader public only through Edward Snowden's revelations on NSA mass surveillance.

to the extent that future scientific breakthroughs will eventually occur outside the existing institutions, in a global open community of researchers who venture into the great unsolved questions of fundamental science.

Whether the intelligence of individual geniuses will play the same role as in the past is another question, however. In their book *At Our Wit's End*, British researchers Woodley and Dutton argue that humanity as a whole has become less intelligent over the past century. Or perhaps in the not-too-distant future, scientists will create artificial brains that are superior to the human mind in every respect.[303] Reconciling such a superintelligence with our values is one of the greatest challenges of the near future, although the general public is barely aware of it yet.

I dread the day when our technology
surpasses our humanity.
- *Albert Einstein*

It was this context that prompted Elon Musk to found his company *NeuraLink*. Its aim is to connect the human brain to a computer, making a potential superintelligence available to everyone before it can be abused by individuals.

Despite these dangers, artificial intelligence has the potential to solve problems that, sooner or later, will push the biosphere to its limits. Climate change is the issue that has gained the greatest attention, but there are many other concerns. There are risks that humans cannot control, but which threaten a technology-dependent civilization as a whole. These include asteroid impacts (probably with warning times during which countermeasures can be developed), supervolcano eruptions, and even geomagnetic storms caused by solar flares, which, in the worst case, could lead to a collapse of the power grids. This would have catastrophic consequences after just a few days.[I]

NOT THE WORST CASE YET

The coronavirus pandemic was only a taster of what can be unleashed by more aggressive or even more contagious viruses. Of course, this also includes biological warfare, which has long been possible thanks to high-tech laboratories[304] in several countries that can manipulate viruses at will. Not to be underestimated are resistant pathogens that can spread via the food chain. The erosion of soils and the pollution of the oceans are of concern, as is the dramatic loss of biodiversity. The reduction of the natural gene pool may reach tipping points as dangerous as those discussed for climate. Perhaps even more insidious than all of the above, are those hazards that Nassim Taleb calls "unknown unknowns," that is, stealthy trends that go unnoticed but undermine the essentials of human life until it is too late.

At the risk of making a somewhat unfair accusation against America, all of these issues, which are seriously underrepresented in everyday politics, are the consequence of a civilization that thinks too short-term: Its cultural basis dates back to about 100 years ago, when the power balance shifted from Europe to the USA.

> The nations of today's world are like a group of climbers connected by a rope. Either they ascend to the summit together, or they fall into the abyss together. – *Mikhail Gorbachev*

[I] A particularly vivid depiction of this can be found in Marc Elsberg's novel *Blackout.*

This development cannot be undone easily; apart from that, even in the heyday of European science, there were no solutions to the global problems of an interconnected world. At least a few basics should continue to apply, however: The primacy of reason, freedom of the mind, truthfulness in the search for knowledge, and control by evidence. By contrast, a worrying irrationality can be observed today; science, which has failed to properly communicate its methods, is not innocent here. Only when researchers once again exhibit an ethos of truth-seeking can they serve as role models for reason- and evidence-based action.

All nations, on the other hand, must understand that for the future prosperity of this civilization gathering on one lonely planet, cooperation is necessary, not aggressive striving for dominance. Whether we can learn this behavior, which is unfortunately not an inbuilt part of human nature, remains to be seen.

Outlook

Quo vadis, Homo sapiens?

As anthropologist Yuval Harari writes in his best-selling book *Homo Sapiens*, the Western world today embraces a culture of consumerism. Possessing and consuming material things is considered a successful way of life. Even the supposedly superior immaterial values, such as experiences, traveling, or recreational activities, arise more from the desire to fill one's life than to give it meaning. Experiences matter more than knowledge. This philosophy of life is certainly a result of the American imprint on modernity.

Even if we assume – either in the hope for a benign superintelligence or as a cautious technology optimist – that the above problems are surmountable, the question remains as to where humanity is heading. At the end of his book, Harari refers to the humans who dominate nature as "dissatisfied, irresponsible gods who have no idea what they want." There is much truth in this. In the superficial culture that currently prevails, 24/7 gratuitous fun for all mankind seems to be the goal – if attainable at all, a somewhat boring paradise. What, on the other hand, is really worth striving for?

Elon Musk, certainly "American" in his style as an entrepreneur, is considering the colonization of outer space. Just as children who want to improve the world should be able to clean up their room, humans must first preserve the conditions for living on their planet, which, incidentally, Musk does have in mind. However, for a meaningful continuation of human endeavor, something more is needed. It is not the striving for power, territory, or experiences that have ultimately taken humanity to new levels, but insights.

This is where what until now may have sounded banal comes into play: a refocusing on European values that is based on still older traditions of the quest for knowledge. Understanding the world has always been superior to the mere improvement of living conditions; the latter was only a consequence of that understanding. At this point, the Confucian quest for wisdom in Chinese culture can contribute more than the imitation of things that have been invented in Europe and developed in the U.S.

The desire to preserve the planet must be accompanied by a genuine interest in the puzzles that lie behind the existence of Homo sapiens in this hostile universe: Are space and time the basis of the real world at all? What are the general laws of nature in that world? Where do constants of nature come from? What characterizes human intelligence? How did life come into being? Do intelligent forms of life exist outside of the Earth?[?] [305]

Intellectual progress of humankind is possible only if we take up the problems that have remained unsolved in fundamental science since 1930. There are plenty of widely unknown ideas of great thinkers that would merit pursuit.[306] Demis Hassabis, founder of the exceptional artificial intelligence (AI) company DeepMind,[307] recently confessed that the motivation for his involvement in AI was to better understand the laws of nature. Whether this can actually be achieved with computers, or should be left to human intelligence, or proves too difficult altogether, we do not know. However, the best we can do to gain the respect of future civilizations is not to give up striving for it.

THANKS

For their inspiring conversations, comments, and corrections, I would like to thank in particular Karl Fabian, Jan Preuss, Freia Unzicker, Wätzold Plaum, and Ernst Peter Fischer. Katharina Unzicker was of great help for the layout.

Literature

Here are also some previously unmentioned books that help to clarify the overall context. Some titles are available in German only. The references for quotes sometimes refer to the German edition.

Andersen, Kurt: Fantasyland – How America went Haywire (Ebury 2017)

Bittner, Wolfgang: *The Conquest of Europe by the USA* (Die Eroberung Europas durch die USA, Westend Verlag 2017)

Bostrom, Nick: *Superintelligence* (Oxford University Press 2014)

Charlton, Bruce G.: *Not Even Trying: The Corruption of Real Science* (2012, Univ. Buckingham press).

Charlton, Bruce G.: *Genius Famine* (Legend Press 2016)

Chomsky, Noam: *Manufacturing Consent* (Vintage 1995)

Chomsky, Noam: Who Rules the World? Reframings (Penguin 2016)

Chomsky, Noam: Requiem for the American Dream (Penguin 2017)

Collins, Harry: *Gravity's Shadow* (Univ. Chicago Press 2004)

Collins, Harry: *Gravity's Ghost* (Univ. Chicago Press 2011)

Clinton, Bill: *Giving* (Cornerstone Digital 2014)

Comay, Ofer: *Science or Fiction?* (Samuel Wachtman's Sons, 2014)

Consa, Oliver (2020): "Something is rotten in the State of QED," https://vixra.org/abs/2002.0011.

Consa, Oliver (2020a): "The Unpublished Feynman Diagram IIc," in *Progress in Physics* 16 no. 3, https://arxiv.org/abs/2010.10345.

Consa, Oliver (2021): *Progress in Physics* 17 no. 9, https://arxiv.org/abs/2109.03301.

Consa, Oliver (2021a), https://arxiv.org/abs/2110.02078.

De Solla Price, Derek: *Little Science, Big Science and beyond* (Columbia University Press 1986).

Davis, P.C.W. and Brown, J.: *Superstrings* (Cambridge University Press 1988).

Douglass, James W.: *JFK and the unspeakable* (Touchstone 2010).

Dutten, Edward and Woodley of Menie, Michael: At Our Wit's End (Societas 2018)

Einstein, Albert: *Relativity* (Grundzüge der Relativitätstheorie, Vieweg 1969)

Einstein, Albert: *The World as I see it* (1934; Mein Weltbild, Ullstein 1988)

Elias, Norbert: On the Process of Civilization (1939)

Farmelo, Graham: *Dirac – The Strangest Man* (Faber and Faber 2009)

Feyerabend, Paul: *Against Method* (Verso 1975)

Feynman, Richard: *Surely You Are Joking, Mr. Feynman!* (W.W. Norton 1985)

Feynman, Richard: *QED - The Strange Theory of Light and Matter* (Princeton University Press 1985).

Fischer, Ernst Peter: Werner Heisenberg, The Oblivious Genius (Das selbstvergessene Genie, Piper 2001).

Gale, George: *Theory of Science* (McGraw-Hill 1979)

Galison, Peter: *How Experiments End* (Univ. Chicago press 1987).

Gamow, George: *Thirty Years that Shook Physics* (Dover 1966, 2012).

Ganser, Daniele: *Illegal Wars* (Orell Füssli 2016)

Ganser, Daniele: *USA: The Ruthless Empire* (Skyhorse 2023)

Graeber, David: *Debt, the first 5000 years* (Goldmann 2013)

Groves, Leslie: *Now it can be told* (Da Capo Press 2009).

Jones, Sheilla: *The Quantum Ten* (Oxford University Press 2008).

Jungk, Robert: *Brighter than A Thousand Suns* (Mariner Books 1970)

Harari, Yuval: Sapiens – A Brief History of Humankind (Harper 2015).

Heisenberg, Werner: Physics and Beyond (Der Teil und das Ganze, Piper 1969)

Horgan, John: *The End of Science* (Little Brown 1996).

Hossenfelder, Sabine: *Lost in Math* (Basic Books 2018)

Huntington, Samuel: *The Clash of Civilizations* (Simon and Schuster 1996)

Hydrick, Carter: *Critical Mass: How Nazi Germany Surrendered Enriched Uranium for the United States' Atomic Bomb* (trine Day, 2016)

Kahnemann, Daniel: *Fast Thinking Slow Thinking* (Siedler 2012)

Kragh, Helge: *Dirac – A Scientific Biography* (Cambridge University Press 2003)

Kuhn, Thomas: *The Structure of Scientific Revolutions* (University of Chicago Press 1962)

Kumar, Manjit: Quantum: Einstein, Bohr and the Great Debate About the Nature of Reality (Icon Books, 2008).

Kurzweil, Ray: The Singularity is Near (2006)

Kurzweil, Ray: *How to Create a Mind* (Viking 2014)

Leggett, Anthony: *Physics* (Birkhäuser 1990).

Lindley, David: *The End of Physics* (Basic Books 1993)

Lindley, David: *Uncertainty* (Anchor Books 2007)

Lippmann, Walter: *Public Opinion* (1922).

López-Corredoira Martín: *The Twilight of Scientific Age* (Brownwalker Press 2013)

McCulloch, Michael: *Physics from the Edge* (World Scientific 2014)

Meyenn, Karl von: *The Great Physicists* (Die großen Physiker ,C.H. Beck 1999)

Moszkowski, Alexander: *Einstein, Insights into his World of Thought* (Einsichten in Einsteins Gedankenwelt 1920, Jazzybee Verlag 2016).

Nachmannsohn, David: German-Jewish Pioneers in Science 1900-1933 (Springer 1979)

Otte, Max: *The Crash is Coming* (Der Crash kommt, Econ 2006)

Otte, Max: *The Long Shadow of Oswald Spengler* (Der lange Schatten Oswald Spenglers, Sonderwege 2018)

Otte, Max (2018): *Financial Markets and the Economic Self-Assertion of Europe* (Die Finanzmärkte und die ökonomische Selbstbehauptung Europas, Springer).

Padderatz, Gerhard: *America – by force into Theocracy* (Amerika - mit Gewalt in den Gottesstaat, Mitteldeutscher Verlag 2007)

Pais, Abraham: *Subtle is the Lord* (Oxford University Press 1982)

Penrose, Roger: The Road to Reality (Vintage 2004)

Pickering, Andrew: *Constructing Quarks* (University of Chicago Press 1984)

Pluckrose, Helen and Lindsay, James: *Cynical Theories* (Swift Press 2020).

Popper, Karl: The Open Society and its Enemies (1945)

Popper, Karl: The Logic of Scientific Discovery (1934)

Popper, Karl: *Objective Knowledge* (Oxford University Press 1972)

Postman, Neil: Technopoly (Vintage 1993)

Rider, Todd H.: Forgotten creators: How German-Speaking Scientists and Engineers Invented the Modern World.

Rosenthal-Schneider, Ilse: *Encounters with Einstein, von Laue and Planck* (Begegnungen mit Einstein, von Laue und Planck, Vieweg 1988)

Russell, Bertrand: Western History of Philosophy (1946)

Sagan, Carl: *Cosmos (*1980)

Sambursky, Samuel: *The Physical World of Late Antiquity* (California University Press 1987)

Sanders, Robert: *The Dark Matter Problem* (Cambridge University Press 2010).

Scahill, Jeremy: Dirty Wars - The World is a Battlefield (Bold Type Books 2013)

Schirach, Richard von: *The Night of Physicists* (House Publishing 2015)

Schrödinger, Erwin: *Nature and the Greeks* (Die Natur und die Griechen, Rowohlt 1956)

Schrödinger, Erwin: *My Life, My World View* (Mein Leben, meine Weltansicht, dtv 2006)

Segrè, Emilio: *From X-Rays to Quarks* (Dover 2007)

Simonyi, Karoly: *A Cultural History of Physics* (Taylor & Francis 2012)

Smolin, Lee: *The Trouble with Physics* (Houghton Mifflin 2006)

Taleb, Nassim: *The Black Swan* (Knaus 2007)

Taleb, Nassim: *Skin in the Game* (Random House 2020)

Taubes, Gary: *Nobel Dreams* (Random House 1987)

Unzicker, Alexander and Jones, Sheilla: *Bankrupting Physics* (Macmillan 2013)

Unzicker, Alexander: *Einstein's Lost Key* (CreateSpace 2015)

Unzicker, Alexander: *The Higgs Fake* (CreateSpace 2013)

Unzicker, Alexander: *The Liquid Sun* (Amazon 2023)

Unzicker, Alexander: *The Mathematical Reality* (Amazon 2019)

Walker, Mark: Nazi Science: *Myth, Truth and the German Atomic Bomb* (Basic Books 2001)

Weinberg, Alvin: *Reflections on Big Science* (MIT Press 1969)

Weinberg, Steven: *The First Three Minutes* (Piper 1997)

Will, Clifford: *Was Einstein Right?* (Basic Books 1986)

Wulf, Andrea: The Invention of Nature and Alexander von Humboldt's New World (Vintage 2016)

Notes

---

1 As recently described by Swiss historian Daniele Ganser (2020).

2 Einstein (1934), p. 43.

3 E.g., Lindley (1993), Horgan (1996), Smolin (2006), Unzicker (2012), Hossenfelder (2018), López-Corredoira (2013).

4 Einstein (1934), p. 105.

5 Cf. Padderatz (2007).

6 Cf. ch. 5.

7 Dirk Hoerder/Diethelm Knauf [eds.] Aufbruch in die Fremde: Europäische Auswanderung nach Übersee (Temmen 1992), p.27 (*European emigration overseas*, German)..

8 Heisenberg (1969), p. 115.

9 Cf. Seymour Martin Lipset: *American Exceptionalism* (New York 1996). See also the book by Andersen (2017): How America went haywire.

10 The books by Lippmann (2018) and Chomsky (1995) are illuminating in this regard.

11 Elias (1939), p. LXXIV (abridged).

12 A refreshing scene about this can be found on YouTube (Jim Bouder): The most honest three and a half minutes of television.

13 Heisenberg (1969), p. 200; however, this is not a literal quotation.

14 S. Sambursky: *Das physikalische Weltbild der Antike* (1965), p. 20.

15 cf. Edward Snowden: *Permanent Record* (2020); Glenn Greenwald: *No place to Hide* (2014)

16 Die Anstalt, German political comedy.

17 Many European physicists, such as Fritz Haber, but also Otto Hahn, also had no qualms about developing weapons such as poison gas.

18 Senate Report, 9/4/1946, https://www.nsf.gov/about/history/EndlessFrontier_w.pdf.

19 See, for example, the 1968 book *The American Challenge* by

former *Express editor* Jean-Jacques Servan-Schreiber.

[20] Cf. Douglass (2010).

[21] Cf. Harari (2015), p. 242; Roland Nelles, "Money, bombs, sense of mission," www.spiegel.de, 07.10.2016.

[22] However, this quote fell after a provocation by a drunken Indian.

[23] Bittner (2017), loc. 1705.

[24] In his book *Cancer as a Metabolic Disease* (Wiley 2012).

[25] For example, Max Otte in his book *Der Crash kommt* (2006) or Nassim Taleb.

[26] Otte (2018), p. 192.

[27] Otte (2018), p. 197.

[28] Schrödinger (1934), p. 29.

[29] Einstein (1934), p. 171.

[30] Cf. Nachmansohn (1979), p.21.

[31] Segrè (1982), p. 293

[32] Wulf (1016), loc. 2254.

[33] Maxwell often referred to Weber, who had developed his own theory; worth mentioning is the research of Andre Koch Torre de Assis.

[34] Autobiographical. Schlipp [5.6] p. 18 (in Simonyi, p. 393).

[35] See, for example, Sir E. Whittaker: *A History of the Theories of Aether and Electricity* (Dover 1951) or A. Unzicker: *Nonlinear continuum mechanics with defects resembles electrodynamics—A comeback of the aether?* (ZAMM 2022)

[36] The equations of relativity even follow from an aether approach, cf. C. F. Frank, Proc. Phys. Soc. A 62 (1949), p. 131.

[37] Simonyi (1978), p. 393.

[38] Lindley (1993), p. 196.

[39] Meyenn (1999), p. 37.

[40] Simonyi (1978), p. 223.

[41] Paul Israel: *Edison, a Life of Invention* (Wiley 2000).

[42] https://de.wikipedia.org/wiki/Nikola_Tesla

[43] Simonyi (1978), p. 393.

[44] There is a historical controversy about who was actually the first; however, there is no reliable evidence to support the opposing view.

[45] For example, in 1915 the Einstein - de Haas experiment that measures the angular momentum of electrons.

[46] Simonyi (1978), p. 407.

[47] With the exception of Bishop Berkley, who had already addressed this in 1721.

48 Simonyi (1978), p. 404.

49 An excellent explanation is *Newton vs. Mach: The Bucket Experiment*, Dialect (YouTube)

50 See also A. Unzicker, *Physics Essays* 34 (2021), 3.

51 Unfortunately, Planck's law is often incorrectly applied to gases today, cf. Robitaille, http://www.ptep-online.com/2008/PP-14-07.PDF.

52 https://www.baslerstadtbuch.ch/stadtbuch/1985/1985_1815.html.

53 Veritasium (Youtube): The man who killed millions and saved billions.

54 Ernst, S: Lise Meitner to Otto Hahn, Wissenschaftliche Verlagsgesellschaft, Darmstadt (1993); Schirrmacher https://www.pro-physik.de/physik-journal/die-physik-im-grossen-krieg. (Physics in the great war (German)).

55 Cf. Landau-Lifshitz, Theoretical Physics II, § 75; Feynman Lectures on Physics II, ch. 28.

56 Heisenberg (1969), p. 79ff.

57 H. Sievers, arxiv.org/abs/physics/9807012, p. 32

58 Through the experiments of Davisson and Germer in 1927.

59 G. Lochak, in S. Diner (ed.), The Wave-Particle Dualism (Springer Netherlands 1983), p. 1ff.

60 Heisenberg, W.: Annalen der Physik 384 (1926), p. 734.

61 Kragh (2003), p. 71.

62 Kragh (2003), p. 69.

63 W. Pauli, Wissenschaftliche Korrespondenz, vol.2, p. 404.

64 Einstein (1934), p. 17.

65 Cf. Bacciagaluppi, https://arxiv.org/abs/quant-ph/0609184.

66 However, this episode probably took place in the 1930 Solvay Conference.

67 Schrödinger (1956), p.27.

68 Kragh (2003), p.81.

69 Schrödinger (1956), p. 13f.

70 Kragh (2003), p. 171.

71 https://en.wikipedia.org/wiki/ Aspect's_experiment, for which the Nobel prize was awarded in 2022.

72 Cf. Jones (2008); see also Frauchinger and Renner, "Quantum theory cannot consistently describe the use of itself," www.nature.com, Sept. 18, 2018.

73 Heisenberg (1969), p. 115

74 Cf. Heisenberg (1969), p. 116ff, who describes in chapter 8 "Atomic physics and practical thinking" a long dialogue with a

young American experimental physicist named Barton.

[75] Einstein (1934), p. 42.

[76] The Lick, later the Yerkes Observatory, both built by private funds.

[77] E. Schrödinger, Annalen der Physik 382 (1925), p. 325 ff.

[78] Farmelo (1979), p. 301.

[79] Kumar (2008), loc. 2963.

[80] Kragh (2003), p. 170.

[81] Gale (1979), ch.7.

[82] https://pressbooks.pub/simplydirac/chapter/natural-philosopher/

[83] The Chinese Chao Chung-yao played a major role in the idea.

[84] Helge Kragh (Dirac's biographer) in Meyenn (1999), p. 352.

[85] Meyenn (1999), 365.

[86] Cf. Hossenfelder (2018).

[87] Kragh (2003), p. 182.

[88] The classical view „Nuclear electron hypothesis" is discussed by Roger H.Stuewer (Otto Hahn and the rise of nuclear physics, p.19-67).

[89] Einstein (1934), p. 116.

[90] Sitzungsberichte der Preußischen Akademie der Wissenschaften,

[91] Segrè (1982), p. 305.

[92] Leo Szilard had already registered a patent for the cyclotron in Berlin in 1929.

[93] Feynman (1985), p. 129.

[94] Kragh (2003), ch. 8.

[95] Kragh (2008), p. 166.

[96] Kragh (2003), p. 166.

[97] Nachmansohn (1979), p. 9.

[98] Einstein (1934), p. 83.

[99] Nachmansohn (1979), p. 16 cites the number 1150 scientists, from Germany alone.

[100] Nachmansohn (1979), p. 31.

[101] Einstein (1934), p. 81ff.

[102] Kumar (2008), p. 293.

[103] Segrè (1982), p. 241.

[104] Another explanation for this is discussed in Heydrick (2016), cf. ch. 9.

[105] Jungk (1956), p. 42.

[106] Nachmansohn (1979), p. 116.

[107] Jungk (1956), p. 278.

---

[108] Jungk (1956), p. 72.

[109] Jungk (1956), p. 74. He refers to the experiments in Rome in 1934, which had been misinterpreted - perhaps to the benefit of mankind (retranslated).

[110] Jungk (1956), p. 85; Unzicker (2012), p. 35.

[111] Jungk (1956), p. 87.

[112] Hydrick (2016), loc. 1443.

[113] Jungk (1956), p. 94f.

[114] Jungk (1956), p. 95.

[115] Known by the code name MAUD, see Schirach (2012), p. 112; Farmelo (2009), p. 306.

[116] Kumar (2008), p. 322.

[117] Ralf Georg: "Weizsäcker's Atomic Bomb Patent", www.welt.de, 20.03.2005; https://www.nbarchive.dk/collections/bohr-heisenberg/documents/; https://physicstoday.scitation.org/doi/10.1063/1.1472389

[118] Heisenberg (1969), p. 212ff.

[119] https://digital.deutsches-museum.de/item/FA-002-533/

[120] https://digital.deutsches-museum.de/item/FA-002-736/#0149, cf. https://en.wikipedia.org/wiki/Calutron.

[121] https://www.archiv-berlin.mpg.de/81907/quelleninventar;

[122] Schirach (2012), p. 105.

[123] This plant used more electricity than all of Berlin, but never produced this plastic, see Hydrick (2016), loc. 1646.

[124] Ganser (2020), ch. 7

[125] Jungk (1956), p. 202.

[126] US National Archives II, *Manifest of Cargo for Tokyo on Board U-234,* RG 38-370 15/05/07 box 3, see Hydrick (2016), loc. 705, loc. 780.

[127] Jungk (1956), p. 198.

[128] Jungk (1956), p. 207.

[129] Jungk (1956), p.228.

[130] https://origins.osu.edu/history-news/hiroshima-military-voices-dissent

[131] http://blog.nuclearsecrecy.com/2016/09/30/fdr-and-the-bomb/

[132] More than against Japan, the bombs were probably directed against Stalin, whose invasion of Japan was feared. The Pacific would then no longer have been the sole sphere of influence of the USA.

[133] James Gleick: Genius - The Life and Science of Richard

Feynman (Vintage 1993), p. 232.

[134] Vasily Arkhipov and Stanislav Petrov, who in 1961 and 1983, respectively, prevented nuclear escalation, deserve special mention here.

[135] The books by Pickering (1984), Taubes (1987), Galison (1987), and also Collins (2004) are very informative in this regard.

[136] Charlton (2012), loc. 1334.

[137] Kragh (2003), p. 185.

[138] Kragh (2003), p. 177.

[139] Jungk (1956), p. 280.

[140] Kragh (2003), p.166f.

[141] Kragh (2003), p.184.

[142] According to nuclear physicist Jörn Bleck-Neuhaus (Elementare Teilchen, Springer, 2012, p. 691), this argumentation is largely based on circular reasoning anyway.

[143] Consa (2020,2020a, 2021).

[144] Consa (2020), 3.1.

[145] J. Schwinger, Physical Review 1948. v. 73(4), p.416.

[146] Consa (2020), 4.1.

[147] Physical Review 74 (1948), p. 1439 and 75 (1949), p. 651.

[148] F. Dyson, Physical Review, 1949. v. 75(3), 486-502.

[149] Consa(2020), 4.2.

[150] Consa (2020), 5.1.

[151] Consa (2020a), even more detailed Consa (2021, 2021a).

[152] Consa (2020), 5.4; Consa (2021), p.2.

[153] Consa (2020), 5.3

[154] Pickering (1984), p. 14.

[155] Schupp A.A., Pidd R.W., Crane H.R. Phys. Rev., 1961 v. 121(1), 1-17; Wilkinson D.T., Crane H.R., Phys. Rev., 1963 v. 130(3), 852-863.

[156] Consa (2020), 6.1, Consa (2021a).

[157] B. T. Aoyama et al, arxiv.org/abs/0706.3496.

[158] Cf. Phys. Rev. Lett. 75 (1995), p. 4728; arxiv.org/abs/hep-ph/0210322; Unzicker (2012), p. 124.

[159] Consa (2021a), p.5.

[160] This is called pseudoconvergence, cf. F. Dyson, Phys. Rev. 85 (1952), p. 631.

[161] Consa (2020), 8.1.

[162] Consa (2020), 2.1.

[163] Consa (2020, 2020a, 2021, 2021a).

[164] Sabine Hossenfelder, "Is the Standard Model of Physics Now Broken? " www.scientificamerican.com, 07.04.2021.

165 Smolin (2006), p. xxiii.
166 Consa (2020), 2.2.
167 Allegedly, he had been told this by astronaut Sally Ride.
168 Leggett (1990), p. 227.
169 Davis (1988), p. 193.
170 E.g., the lack of description of the important coincidence count, cf. Unzicker (2012), chap. V-2.
171 Gale (1979), ch.7.
172 Cf. ch. 8; Bohr and Casimir presented Heisenberg's views quite differently from his own (Fischer 2001, p. 167); see also Schirach (2012) and Walker (2001).
173 Ernst Peter Fischer: "The Philosopher Who Loved the Atomic Bomb," www.welt.de, June 26, 2012; This became known only after Russian archives were opened after 2000.
174 https://www.nobelprize.org/prizes/physics/1968/summary/
175 Gamow (1966), ch. IX.
176 E. Fermi, Phys. Rev. 76 (1949), p. 1739.
177 Gamow (1966), p. 173.
178 Cf. Landau-Lifschitz, Theoretical Physics II, § 75; Feynman Lectures II, ch. 28.
179 Nominations become public after about 50 years, https://www.nobelprize.org/nomination/physics/
180 Pickering (1984), p. 12.
181 Segrè (1982), p. 297.
182 Pickering (1984), p. 143.
183 Cf. Heisenberg (1969), p. 123)
184 Unzicker (2012), p. 181.
185 Cf. Kahnemann (2012).
186 J. Ralston, Arxiv.org/abs/1006.5255.
187 Charles Enz/Karl v. Meyenn: Wolfgang Pauli - das Gewissen der Physik (Springer 1988), p. 450.
188 Segrè (1982), p. 293.
189 Lindley (1993), p. 167.
190 Pickering (1984), p. 111.
191 Pickering (1984), p. 14.
192 Taubes (1987), p. 78.
193 Einstein (1934), p. 105.
194 Weinberg (1969), p. vi.
195 Einstein (1934), p. 172.
196 Pais (1982), loc. 83.
197 Pais (1982), loc. 203.

198 A. Einstein, Annalen der Physik 35 (1911), p. 906, www.physik.uni-augsburg.de/annalen.

199 The translation of Einstein's complete works (Einstein Papers Project) progressed very slowly and has only recently been completed.

200 Cf. Will (1986)

201 Cf. Galina Weinstein, https://arxiv.org/abs/1602.04040.

202 Cf. Collins (2004).

203 Cf. Collins (2004), p. 395, retranslated from German.

204 Cf. Cóllins (2004), p. 372.

205 Cf. A.D. Jackson, https://arxiv.org/abs/1706.04191; S. Hossenfelder, "Missing Link: Doubts Despite Nobel Prize - Dispute Over Gravitational Wave Measurements," www.heise.de, 27 Oct. 2019. Suspension as possible source of error: https://arxiv.org/abs/2009.00400.

206 Jeremy Bernstein, "Albert Einstein and Black Holes," www.spektrum.de, Aug. 01, 1996 (magazine).

207 W. Kundt, J. Astrophys. Astron. 10 (1989), p. 119; J. Casares, http://www.arxiv.org/abs/astro-ph/0612312.

208 Claims of a Black Hole Image: the Day Astrophysics Died, SkyScholar (YouTube)

209 Schrödinger (1934), p. 118.

210 https://en.wikipedia.org/wiki/We_choose_to_go_to_the_Moon

211 https://en.wikipedia.org/wiki/List_of_German_aerospace_engineers_in_the_United_States, see also Rider (2017).

212 Discussed in detail in Robitaille (2011), http://fs.unm.edu/PiP-2011-03.pdf.

213 NASA SDO - Spectacular Prominence Eruption, www.youtube.com, 07.06.2011.

214 Rollin Thomas Chamberlin of the University of Chicago.

215 Acronym for *Mathematical Analyzer Numerical Integrator and Automatic Computer Model.*

216 https://de.wikipedia.org/wiki/Julius_Edgar_Lilienfeld

217 Segrè (1982), p. 295.

218 Justifiably criticized by Sabine Hossenfelder, "The World Doesn't Need a New Gigantic Particle Collider," www.scientificamerican.com, 6/19/20.

219 R. Pohl et al, Nature 466 (2010), p. 213.

220 The latter was the basis of the 1979 Nobel Prize. For an illuminating analysis of these constructs, see Pickering (1984); see also Comay (2014).

---

221 For example also Einstein in his article Mathematische Annalen 102 (1930), p. 685, or Dirac.

222 Taubes (1987), p. 6f.; p. 135.

223 Taubes (1987), p. 6.

224 Cf. e.g. Unzicker, A.: http://www.arXiv.org/gr-qc/abs/9612061, sec. 4.

225 Ryder, L.: Quantum Field Theory (Cambridge 1996), p. 2: "For these reasons, the particle physicist is justified in ignoring gravity - and is glad to do so!"

226 However, Robitaille (2009) raises legitimate objections to this interpretation of the data: http://www.ptep-online.com/2009/PP-19-03.PDF

227 M. Disney, https://arxiv.org/abs/astro-ph/0009020.pdf.

228 Jones (2010), p. 7.

229 E. Schrödinger, Annalen der Physik 382 (1925), p. 325 ff.

230 Penrose (2004), p.753.

231 Simonyi (1978), p. 431.

232 The so-called torsion. Cf. Annalen der Mathematik 102 (1930), p. 685; R. Debever: *Einstein-Cartan Letters on Absolute Parallelism 1929-1932 (*Princeton University Press 1979).

233 Smolin (2006), pp. xxii, xxiii, viii.

234 https://www.math.columbia.edu/~woit/wordpress/?p=4696

235 Hossenfelder (2018), loc. 1339.

236 Feynman (1985), p. 150.

237 Sanders (2010), p. 115ff.

238 P. Kroupa, https://arxiv.org/abs/1204.2546.

239 Lindley (1993), p. 222.

240 https://history.fnal.gov/criers/FN_1998-01-23.pdf.

241 YouTube: Fox news speaks with Michio Kaku about the LHC

242 Smolin (2006), p. xv.

243 F. Wilczek, arXiv.org/abs/0708.4361, p. 19.

244 Smolin (2006), p. xvii.

245 For example, in the conference "Why trust a theory?" 2015 in Munich.

246 Heisenberg (1969), p. 36.

247 John Horgan's interview in *The End of Science* (1996) is also informative.

248 J. Horgan, "Physics Titan Still Thinks String Theory Is 'On the Right Track,'" blogs. scientificamerican.com, Nov. 22, 2014.

249 David Gross, one of the most staunch advocates of the theory, is shown to grossly misrepresent key statements of Dirac's

paper. "Real Physics Talk - David Gross." www.youtube.com, 10.08.2016.

250 Charlton (2012), loc. 450.

251 Worth seeing: "Karl Popper - A Conversation (1974)", www.youtube.com, 04.07.2013.

252 Replication crisis" entry on en.wikipedia.org

253 Hossenfelder (2018), loc. 1141.

254 "The Morality of Fundamental Physics - Nima Arkani-Hamed - Cornell UniversityChannel," www. youtube.com, 15 Apr 2019.

255 YouTube (zehadi Alam), Brian Greene v. Neil deGrasse Tyson; Hossenfelder (2018), loc. 1349.

256 https://videoonline.edu.lmu.de/en/node/7478

257 Cf. Lindley (1993), p. 255.

258 Just think of cum-ex and cum-cum.

259 Charlton (2012), loc. 183.

260 S. Hossenfelder; "The World Doesn't Need a New Gigantic Particle Collider." www.scientificamerican.com, 19.06.2020.

261 YouTube (Unzicker's Real Physics): Anne L'Huillier (Nobel Committee), t=598s.

262 A. Unzicker: "Physics Nobel Prize for Secret Science," www.heise.de, 07.10.2015.

263 Cf. ch. 10.

264 For example, a group from Copenhagen has uncovered serious flaws in the LIGO data analysis on gravitational waves, cf. https://arxiv.org/abs/1706.04191; A. Unzicker: "Gravitationswellen: Silent Fiasco," www.heise.de, Feb. 16, 2020; S. Hossenfelder, "What's up with LIGO? ", backreaction.blogspot.com, 04.09.2019.

265 R. Cowen, "Gravitational Waves Discovery Now Officially Dead." www.scientificamerican.com, 02.02.2015.

266 Regardless, there are reasonable doubts about this, see http://www.ptep-online.com/2009/PP-19-03.PDF.

267 Charlton (2012), loc. 461.

268 Cf. A. Unzicker: "Dark Matter as a Permanent Surprise," www.heise.de, 30.11.2014.

269 Hossenfelder (2018), loc. 3158.

270 Cf. Unzicker (2012), p.51.

271 https://www.math.columbia.edu/~woit/wordpress/?p=4262

272 This is also supported by the interviews in John Horgan's book (1996), p.271.

273 De Solla Price (1986), p. xix.

274 Cf. Charlton (2016).

275 Smolin (2006), p. 347.

276 Worth mentioning here is the *Declaration of Academic Freedom*, http://www.ptep-online.com/2006/PP-04-10.PDF.

277 Charlton (2012), loc. 470.

278 Charlton (2012), loc.1511.

279 Einstein (1934), p. 23.

280 Einstein (1934), p. 23.

281 Smolin (2006), ch.5.

282 Hossenfelder(2018), loc. 2520; www.nature.com/articles/d41586-021-01436-7

283 Charlton (2012), loc. 1542.

284 Davide Castelvecchi, "CERN suspends physicist over remarks on gender bias," 01.10.2018.

285 https://philarchive.org/rec/CORFIS-2; Reflections On The Statement "The IAU Has A Clear Ideology About Inclusion That Has To Be Accepted By All Its Members" (www.science20.com).

286 en.wikipedia.org/wiki/Google%27s_Ideological_Echo_Chamber

287 Kevin Reed: Science professor's MIT lecture canceled over views on affirmative action, wsws.org, 10/22/21.

288 What should one say, for example, about a professorship of gender-equitable language in civil engineering at the TU Aachen?

289 China Becomes Leader in Science as U.S. Schools Prioritize "Diversity." P. Deift, S. Jitomirskaya, S. Klainerman, "As US Schools Prioritize Diversity Over Merit, China Is Becoming the World's STEM Leader," quillette.com, 8/19/2021; Lawrence Krauss, "How 'Diversity' Turned Tyrannical," www.wsj.com, 10/21/21.

290 Google Academy, https://learndigital.withgoogle.com/.

291 Bruce G. Charlton, Medical Hypotheses 72 (2009) 237-243.

292 E.g., Ganser (2020) ch. 2; loc. 660

293 It is quite clear that this debt cannot be repaid; cf. Graeber (2013).

294 https://caityjohnstone.medium.com/its-not-okay-for-grown-adults-to-say-the-ukraine-invasion-was-unprovoked-53b7990576c1

295 Focus, 14.08.20: China and Russia prefer to pay each other in euros.

296 Richard T. Cooper : General Casts War in Religious Terms, www.latimes.com, Oct. 16, 2003:

[297] Cf. Ganser (2016).

[298] https://en.wikipedia.org/wiki/Military_budget

[299] Cf. Scahill (2013).

[300] https://www.moonofalabama.org/2021/08/hypersonic-missiles-are-they-a-gamechanger-by-gordog.html#more

[301] de.wikipedia.org/wiki/Circular_Electron_Positron_Collider

[302] Cf. Chomsky (1995); Lippmann (2018). Only a few defenders of free speech, such as Glenn Greenwald, Joe Rogen, or occasionally Tucker Carlson, still offer hope in this regard.

[303] Cf. Bostrom (2016), Kurzweil (2006, 2014).

[304] Wuhan (China) is well known, but the U.S. also operates several labs in Ukraine and Georgia, among others, J. Mitschka, "Is the U.S. Developing New Biological Weapons?", www.heise.de, 28 Sept. 2018; "What Are U.S. Biological Labs Researching in Ukraine?", www.anti-spiegel.ru, 08 Apr. 2021.

[305] A not so far-fetched thesis, https://www.sciencedirect.com/science/article/pii/S0079610718300798

[306] To mention here would be - from my point of view - ideas of Einstein 1911, Schrödinger 1925 or Dirac 1938, cf. Unzicker (2015, 2019).

[307] DeepMind's *AlphaGo* program defeated the world's best Go player in 2016, a feat that most experts at the time had thought impossible.

Made in the USA
Middletown, DE
30 January 2023